Lecture Notes in Chemistry

49

E. Roduner

The Positive Muon
as a Probe
in Free Radical Chemistry

Potential and Limitations of the μSR Techniques

Springer-Verlag

Berlin Heidelberg New York London Paris Tokyo

Author

E. Roduner
Physikalisch-Chemisches Institut der Universität Zürich
Winterthurerstr. 190, CH-8057 Zürich

ISBN 978-3-540-50021-6   ISBN 978-3-642-51720-4 (eBook)
DOI 10.1007/978-3-642-51720-4

2151/3140-543210

Preface

The work presented here is a result of an extended collaboration with a number of coworkers and guests. Particularly, I would like to thank Dr. P. Burkhard and Dr. W. Strub for their careful work performed for their Ph.D. thesis and Dr. M. Heming for his brilliant ideas and his dedication. Very fruitful and stimulating were collaborations with our guests, i.e. with G.A. Brinkman and P.W.F. Louwrier from NIKHEF-K in Amsterdam, B.C. Webster, M.J. Ramos and D. McKenna from the University of Glasgow, M.C.R. Symons, D. Geeson and C.J. Rhodes from the University of Leicester, S.F.J. Cox and C.A. Scott from the Rutherford Appleton Laboratory in Chilton, and R. De Renzi and M. Riccò from the University of Parma. Many invaluable discussions with friends and competitors in the field helped to address new viewpoints and to define new goals. I shall not forget my teacher and director of the radical chemistry group, Prof. H. Fischer, whom I wish to thank for his interest and active support and for the great liberty he allowed me for the planning and organization of the project. Last but not least, I thank my dear wife Hanny and our children Christian, Martin and Andrea who suffered, without complaint, daddy's absence for so many hours.

Financial support by the Swiss National Science Foundation and the excellent experimental conditions at the Paul Scherrer Institute (formerly Swiss Institute for Nuclear Research) are gratefully acknowledged.

A very special thank is due to the Swiss Chemical Society for awarding the author with the *Werner Price and Medal* for the year 1988 for the work on which this article is based.

Contents

Chapter 1

Introduction

1.1 Motivation

The principle motivation of natural sciences is the desire to learn about the nature of matter and about its interaction with forces, with the main aim being to understand fundamentally where we all come from and what it is that keeps nature going. In this century, scientists have not only set off to investigate the macroscopic world of the universe as a whole, they have also succeeded in penetrating the microscopic world to learn more about the structure of molecules, of atoms and their nuclei, and they even break up nuclei to find the most elementary building blocks of matter. A major breakthrough for a detailed understanding of the microworld occurred in the twenties when the concepts of quantum mechanics were introduced. They mark the beginning of the victorious evolution of spectroscopic techniques, which are today the most widespread tools to investigate the microworld. Thereby, matter is subjected to electromagnetic radiation spanning a range of more than ten orders of magnitude in wavelength between X-rays and radio waves. The response of matter to these waves is often very specific, so that it allows detailed insight into its structure and dynamics.

Experimental techniques using radiation are not restricted to mass-less photons. Particles have a wave character as well, and the lighter they are the more noticable it becomes. Irradiation of matter with particles, in particular with electrons and neutrons, for structure determination is common, and the diffraction pattern obtained e.g. by irradiation of a crystal with electrons is not fundamentally different from the diagram resulting from irradiation with X-rays.

Experiments are also not restricted to the study of unperturbed matter. Nature is very complex and non-ideal. It is therefore often the purpose of irradiation to exert a perturbation or induce a specific defect in order to simulate a certain aspect of nature in a well characterized environment. Again, this can be achieved with photons or with particles, and the choice of the technique depends mostly on the kind of perturbation and on the material to be studied. Furthermore we note that a particle can create defects in passing by, or that it essentially constitutes the defect and serves as a perturbing probe in matter. It is this last category which is of interest here. The probe is the positive muon, and the material is investigated in view of its chemical properties.

Chemists are familiar with the proton and the neutron as the fundamental components of the relatively heavy positively charged nuclei, and with the much lighter negatively charged electrons which surround the nuclei in chemical structures. They are much less familiar with other fundamental particles such as muons, pions, kaons and hyperons. Some of them are present in nature, e.g. in cosmic rays, but they decay rapidly and are thus not suited for the synthesis of stable molecules. On the other hand, many of them carry a positive or negative elementary charge. They are thus subject to the Coulomb force which governs chemistry, and one can imagine that for a short time they can take the role of nuclei or of electrons in molecules. Furthermore, some of them have magnetic moments and are therefore capable of sensing magnetic interactions, which are the basis for magnetic resonance, one of the most powerful analytical techniques in chemistry. Among these exotic particles it is the muon in particular which has been used extensively as a probe for magnetic and electronic interactions in matter in the past decade [1-7].

This work was initiated in 1978 with the first direct observation of organic radicals with incorporated muons by the muon spin rotation technique [8]. A logical consequense of this discovery was the aim to explore the potential of the positive muon as a probe in the chemistry of organic free radicals. The technique has been applied to the investigation of various chemical systems in solid, liquid and gaseous states, and with respect to questions of radiation chemical, structural and kinetic interest. A large amount of information has been accumulated. Here, the main results are collected and exemplified using a limited selection of chemical systems, mostly cyclohexadienyl type radicals.

1.2 History and properties of the muon and its bound states

The muon was the first unstable elementary particle observed. It was discovered in 1937 by *Neddermeyer and Anderson*[9] by its traces in photographic emulsions exposed to cosmic rays. It occurs in two charge states, as μ^+ and μ^-, and has a rest mass equal to one-ninth the mass of a proton, or 207 times the mass of an electron. A negative muon can replace an electron in an atom, but its huge mass leads to very low lying orbits and therefore to an atom with quite different chemical properties. We shall confine ourselves to the positive muon which appears to be a useful analogue of the proton. Its physical properties are given in Table 1.1.

It was not until 1957 that scientists realized that the positive muon could form a bound state with a negative electron. This one-electron atom was dubbed muonium [10]($\mathrm{Mu} \equiv \mu^+ e^-$, the suffix *onium* is misleading since it is reserved in physics for particle-antiparticle pairs, and in chemistry for certain cations). Chemically, Mu is a light isotope of hydrogen with a mass equal to one-ninth the mass of H [5]. Its ionization potential and its Bohr radius are, to within 0.5%, the same as those of H, D, and T (see Table 1.2).

Table 1.1: Properties of μ^+

Mass	m_μ	$=$	$0.1126 \cdot m_p = 206.7684 \cdot m_e$
Mean lifetime	τ_μ	$=$	$2.197134 \mu s$
Spin	I	$=$	$\frac{1}{2}$
Magnetic moment	μ_μ	$=$	$3.183344 \cdot \mu_p$
Larmor precession frequency	ν_μ	$=$	$13.5538 \text{ kHz/Gauss}$

Table 1.2: Physical properties of hydrogen isotopes

Property	Mu	H	D	T
Mass ($/m_H$)	0.1131	1.0000	1.998	2.993
Reduced mass ($/m_e$)	0.9952	0.9995	0.9997	0.9998
Ionization potential ($/eV$)	13.539	13.598	13.601	13.602
Bohr radius ($/pm$)	53.17	52.94	52.93	52.93
Hyperfine coupling constant ($/MHz$)	4463	1420	218	1516

It was *Brodskii* [11] who suggested in 1963 that muonated radicals should be formed by addition of Mu to unsaturated molecules such as ethylene. There was indirect evidence from several experiments that such species are indeed formed [12], but a breakthrough occured only in 1978 with their first direct observation in high magnetic fields [8].

1.3   The muon as a probe in matter

1.3.1   Principle of the experiment

Investigation of matter using positive muons involves irradiation of the experimental samples with these particles which are available as energetic spin polarized beams at the ports of special accelerators. The experimental technique of muon spin rotation (μSR) monitors the time evolution of muon spin polarization as a function of external and/or internal magnetic fields. This allows the experimenter to distinguish between signals arising from muons in diamagnetic environments, in Mu or in muonated free radicals, and to follow their interconversion. The details of the technique are described in Chapter 2.

1.3.2   Solid state physics

In many solids the muon comes to rest at a certain position in the lattice. It is then used as a magnetic probe (a 'microscopic Gaussmeter' [7]) in order to measure local fields in ferromagnets and antiferromagnets, Knight shifts in conducting materials, and field distributions in dilute magnetic alloys or in spin glasses. In other solids the muon diffuses untill it finds an impurity or defect where it may get trapped. The local field then changes in time, which results in a temperature dependent relaxation of the signal. Diffusion and annealing behaviour is studied. In insulators and semiconductors isotropic muonium states are observed. Sometimes, one finds a hyperfine interaction which is drastically reduced compared with its value for free Mu. The most interesting situation which has just started to be understood arises from the observation of anisotropic muonium-like states in diamond, silicon and germanium. They have no analogue with hydrogen atoms yet [1,4,6,7].

4

1.3.3　Chemistry

Distribution of muons between chemical states

The distribution process of the muons between the different states usually occurs prior to the direct observation of the species (see Figure 1.1). In the sample, the muon thermalizes within a few centimeters from a kinetic energy of typically 20 MeV, mostly via ionization and excitation of the medium. The time scale and nature of the processes governing the slowing down and the distribution of the muons are of fundamental interest and have been subject to many discussions. Models including hot atom or radiation chemical events have been set up. The hot model has some support from gas phase results [13]. In condensed phases, radiolytical processes are of greater relevance, but a proper distinction with hot processes is often impossible, if not semantic [2,3]. To some extent, the models can be tested indirectly by looking at the response of the observed polarizations when magnetic fields or fast chemical reactions interfere with the time evolution of polarization or with the distribution processes at early times.

Muons in diamagnetic environments

It was already realized in the first experiments that the fraction of muons found in diamagnetic environments depended strongly on the chemical composition of the material investigated [14]. It was usually assumed that these muons are substituted in molecules rather than free. They will be denoted D in general, or $Mu^+ \equiv \mu^+$ if they are not chemically bound.

Molecules labelled with radioactive isotopes are of importance in biological systems. They are chemically indistinguishable from normal molecules and therefore used as tracers to follow metabolitic pathways in intact bodies or for diagnostic purposes in medicine. The short lifetime of the muon does not permit such applications.

Muonium

Being a light isotope of hydrogen, Mu is expected to undergo well-known chemical reactions, e.g.:

- abstraction,

$$Mu + XR \longrightarrow MuX + \cdot R \qquad (1.1)$$

- electron transfer,

$$Mu + M^{n+} \longrightarrow \mu^+ + M^{(n-1)+} \qquad (1.2)$$

5

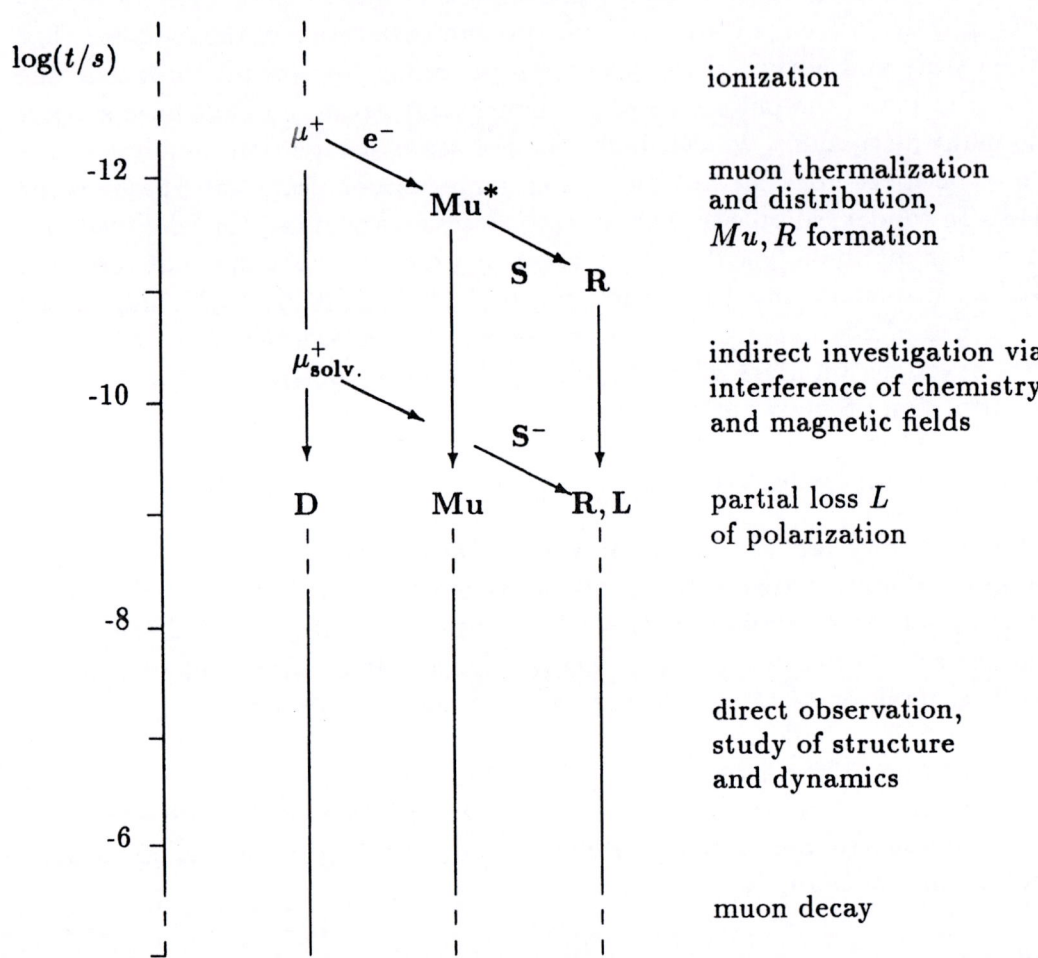

Figure 1.1: Scheme for the formation of muonated species.

- radical combination,

$$Mu + \cdot OH \longrightarrow MuOH \qquad (1.3)$$

- spin exchange,

$$Mu(\uparrow) + Ni^{2+}(\downarrow) \longrightarrow Mu(\downarrow) + Ni^{2+}(\uparrow) \qquad (1.4)$$

- addition,

$$Mu + C_6H_6 \longrightarrow \cdot C_6H_6Mu \qquad (1.5)$$

The relevance of Mu in chemistry is based on its character as a hydrogen isotope. Rate constants for its chemical reactions have been determined and compared with corresponding values for the reactions of H. This reveals kinetic isotope effects of up to two orders of magnitude and more, in both directions [5,15-18]. In abstractions (eqn. 1.1) Mu reacts often much more slowly than H. This is ascribed to the higher zero-point vibrational energy of the transition state involving the lighter isotope [16]. On the other hand, it is often thought that quantum mechanical tunneling contributes to the reactions of H atoms [19]. Such effects should be much more important for Mu and favour, of course, the lighter isotope. Model calculations support this view, and they reveal a strong dependence on barrier height and width. For relatively slow reactions the barriers are high, and the zero point vibrational term dominates the kinetic isotope effect. For faster reactions the barriers are lower and often narrower, and the tunneling term dominates. This was found to be the case in Mu vs. H addition to unsaturated compounds (eqn. 1.5) [20].

Of particular interest are simple reactions in the gas phase which can be treated theoretically. The most prominent example is the reaction of Mu(H,D) with $H_2(D_2)$ [21]. The large kinetic isotope effects observed impose a sensitive test on theories of reaction rates and on the calculated energy hypersurfaces [22].

In reactions (1.1-1.3) the muon ends up in diamagnetic environments, either free or chemically bound in a molecule. In reaction (1.4) it does not change its chemical state, and in (1.5) it becomes incorporated in a paramagnetic species, here the muonated cyclohexadienyl radical. Formally, all muonated radicals R are derived by Mu addition to unsaturated parent molecules, as in (1.5). However, as it will be seen in Section 9.6, the physical formation process does not have to involve Mu. There are cases where the addition of μ^+ and e^- is not concerted.

7

Muonated free radicals

Having accepted that Mu is a light hydrogen isotope one may think about muonated radicals in the same way as about deuterated or tritium labelled radicals. In conventional radical chemistry, specific isotopic substitution of parent molecules is used in mechanistic studies or to aid the assignment of coupling constants [23]. Isotopically enriched samples, in particular with ^{13}C, are used to obtain more detailed information on the electron spin density distribution in structural studies. Investigations of the coupling constants of deuterated methyl and ethyl radicals and comparison with their H analogues revealed deviations which are beyond the ratio of magnetic moments. They have been explained in terms of dynamic effects [24,25]. The vibrational wave functions are mass dependent. This leads to isotope effects in vibrationally averaged coupling constants. Corresponding changes are expected to be much larger in muonated radicals.

Kinetic isotope effects were found for reactions where the bond to the isotope was directly involved, as for example in H/D transfer [19]. Again, differences could be much larger in the muonated case.

In this work, the versatility of the muon as a probe in free radical chemistry, i.e. the formation, identification and transformation of muonated radicals compared to their 'normal' protiated analogues, is investigated.

Chapter 2

Experiments employing muons

2.1 Muon production and decay

'Natural' muons are produced in the following way: light nuclei, mainly protons, from primary cosmic rays fall on the earth with high energy. In collisions with molecules in the upper atmosphere they interact with their nuclei and trigger the production of new particles. One of them, the positive pion, decays with a mean lifetime of 26 ns into a positive muon and a neutrino:

$$\pi^+ \longrightarrow \mu^+ + \nu_\mu. \tag{2.1}$$

The pion is a spin-zero particle, but the neutrino has spin $\frac{1}{2}$ with negative helicity, i.e. with its spin pointing in the direction opposite to its momentum in the centre-of-mass system. In order to conserve angular momentum the muon must also have spin $\frac{1}{2}$ and negative helicity.

In the laboratory, muons are produced in the same way. Protons or electrons are accelerated to an energy exceeding the pion rest mass (0.15 amu or 140 MeV) and collide with a pion production target. Emitted pions with the desired charge and momentum are selected using a dipole magnet. They decay in flight over a distance of several meters. Because of the above characteristic of the pion decay, momentum selection of the muons yields a *spin polarized muon beam*. Alternatively, muons derived from pions decaying at rest at the surface of the production target can be selected. This gives rise to a beam of polarized muons with an energy of 4.1 MeV and a momentum of 29.8 MeV/c. Because of their birth at the target surface they are called *surface muons*.

The muon decays with a lifetime of 2.2 μs into a positron and two neutrinos:

$$\mu^+ \longrightarrow e^+ + \bar{\nu}_\mu + \nu_e. \tag{2.2}$$

As a consequence of the conservation of energy, momentum and angular momentum in this three body decay, the positron is emitted in a direction with an angle θ to the muon spin direction, with a probability proportional to $1 + a \cdot \cos\theta$, i.e. *preferentially along the muon spin direction*. The asymmetry a should equal 0.33 for a fully polarized muon beam, but due to a number of experimental imperfections it is usually 0.2-0.3 [1].

Accelerators producing muons for application in chemistry are found at TRIUMF (Vancouver), KEK (Tokyo), and at the Swiss Institute for Nuclear Research (SIN) in Villigen, where the work presented here was performed. This year, a new intense muon source with a pulsed structure has started to operate at the Rutherford Appleton Laboratory (RAL) in England.

2.2 The μSR techniques

Several variants of experimental techniques have been developped. They have in common that they all stop muons from a spin polarized beam in the sample to be investigated, which is placed in an external magnetic field B. Second, they all monitor the evolution of muon spin polarization in the sample via detection of the decay positrons, utilizing a single particle counting technique which was originally developped by *Garwin et al.* [14] for an experiment verifying parity violation in pion and muon decay. The techniques were dubbed μSR, where μS stands for muon spin and R for rotation, relaxation or resonance, depending on the details of the technique [1-4,7,26]. For chemical applications, the experimental samples are liquid solutions purified from oxygen via freeze-pump-thaw cycles and sealed in spherical glass bulbs of 25-50 mm diameter. Gases are kept in a container with a thin window, and low momentum muons ('surface' muons) have to be used to allow them to stop in low density material.

2.2.1 Time differential transverse field μSR

This is at present the most common technique of μSR, and most results in this work were obtained this way. The sample is placed in the centre of a pair of Helmholtz coils producing a magnetic field transverse to the muon spin polarization (left of Figure 2.1). Scintillation counters in the backward (b) and forward (f) direction with respect to the muon beam detect incoming muons and corresponding decay positrons, and a fast clock measures the time interval between these two events, i.e. the lifetime of individual muons in the sample, with a resolution of a nanosecond. A fast electronic logic discriminates between good and bad events. Cumulative muon stops are not

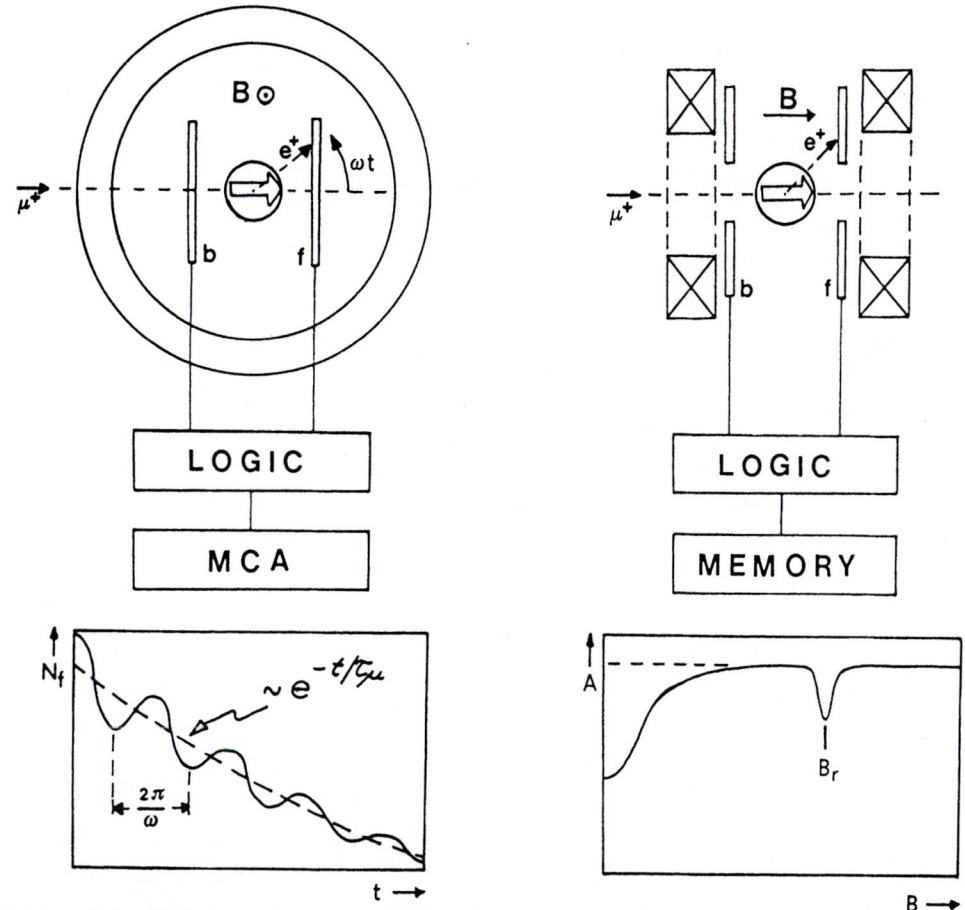

Figure 2.1: Block scheme of μSR apparatus for time differential measurements in transverse fields (left) and for time integrated measurements in longitudinal fields (right).

accepted. The number N of decay positrons as a function of corresponding muon lifetimes is stored in a histogramming memory. For an unpolarized beam this is a simple exponential decaying with τ_μ. For a polarized beam spin precession results in a modulation of the count rate in the positron detector, owing to the anisotropy of the decay. The corresponding oscillations in the histogram represent a free induction decay of the muon polarization, in complete analogy to the FID obtained in FT-NMR after a 90° pulse for a spin-$\frac{1}{2}$ nucleus. Frequency, initial amplitude and phase, and relaxation of this signal contain the information of interest. As in NMR, the signals are interpreted as transitions between magnetic states. They follow specific spectroscopic selection rules.

Table 2.1: Comparison of μSR and FT-NMR

	μSR	FT-NMR
time resolution	1 ns	1μs
frequency resolution	0.5 MHz	0.1 Hz
polarization	0.7	10^{-5}
minimum number of spins for simple spectrum	10^7	10^{17}
detection method	single particle counting	induction coils

Comparison of typical values for important parameters reveal essential differeces between μSR and FT-NMR (Table 2.1). The time resolution in NMR is limited by the duration of a pulse to tip the magnetization angle. This pulse is not needed in μSR since the muon is injected with its spin transverse to the field. The frequency resolution is related to the duration of the FID. It is limited in μSR by the muon lifetime. NMR usually has to work with Boltzman population of the energy levels. This results in a small and temperature dependent polarization. The corresponding property in μSR is given by the polarization in the muon beam. Accordingly, far less spins are needed in μSR than in NMR to produce a detectable signal. This effect is even enhanced by the sensitivity of the single particle detection method. Pulsed techniques have also been developped for ESR of free radicals. Owing to the larger magnetic moment of the electron as compared with the proton the corresponding entries in Table 2.1 for pulsed ESR would usually take values intermediate between those for μSR and NMR.

2.2.2 Time integrated longitudinal field μSR

An alternative technique places the sample in a longitudinal field (right of Figure 2.1). It involves simple counting of decay positrons in the forward (N_f) and in the backward (N_b) direction. The asymmetry,

$$A = (N_f - N_b)/(N_f + N_b), \qquad (2.3)$$

is plotted as a function of field strength. For non-interacting muons, or under spin decoupling conditions (high fields), there is no evolution of muon polarization, and A is constant. For paramagnetic states in low fields the muon polarization relaxes owing to muon-electron hyperfine interaction. This causes a decrease of A and gives a relatively non-characteristic 'quenching curve' for multi-spin systems such as organic radicals. The break through of this technique came only recently, when *Abragam* [27] predicted that avoided level crossings (ALCs) in higher fields should have a similar effect. This was verified experimentally [28,29]. Indeed, sharp dips at characteristic 'resonant' fields B_r were observed, and position, amplitude and width of these signals are expected to yield information complementary to that obtained in transverse field μSR.

2.2.3 Other techniques

Several groups have devised specialities in order to adapt to particular problems or to take advantage of specific accelerator properties. The stroboscopic technique tunes the muon precession in transverse fields to the accelerator frequency [4]. In this way, it can accept multiple muon stops, but the experiment is limited to a single frequency at a time. Another promising approach uses an oscillating field B_1 perpendicular to B_0, in analogy to CW magnetic resonance [30]. This is favourable in the context of pulsed muon sources.

Of course, the longitudinal field technique can also be used in time differential mode in order to measure the time dependence of the muon relaxation, or the oscillation of spin polarization between muon and other magnetic nuclei.

2.3 Analysis and interpretation of the data

The general form of a histogram obtained in transverse field μSR is given by

$$H(t) = N_0\{BG + e^{-t/\tau_\mu}[1 + F(t)]\}, \qquad (2.4)$$

where N_0 is a factor depending only on the total number of counts and is roughly equal to the number of counts in the first channel, BG is the background fraction (usually less than 1%), and $F(t)$ reflects the FID. If several muonated species contribute to the signal and/or if one species shows more that one frequency, $F(t)$ is a sum of contributions of the form

$$F_i(t) = A_i \cdot e^{-\lambda_i t} \cos(\omega_i t + \phi_i), \qquad (2.5)$$

each describing a precession on a specific frequency ω_i with its amplitude (asymmetry) A_i, damping constant λ_i and initial phase ϕ_i. The asymmetries A_i depend on the beam polarization P, the asymmetry coefficient a, the solid angle of the scintillation counters, the partitioning of P between different species and frequencies, and on reaction and relaxation rates. λ_i represents reaction or relaxation processes, and ϕ_i depends on experimental factors as well as on reaction or relaxation in a precursor of the observed state.

The asymmetries A_i are converted into absolute *fractional muon polarizations* P_i by calibration against a standard. Here, it was assumed that the amplitude observed with liquid carbon tetrachloride for muons in diamagnetic environments represents 100% polarization. The accuracy of this assumption is currently a matter of debate [31,32], but for most purposes it is not of importance. Observed polarizations have to be corrected for a number of instrumental factors, i.e. for the effect of the limited time resolution, for muons stopping in the glass container or in the cryostat, and for the effect of absorbed decay positrons as a function of target size and density [33]. Depending on the chemical environments the fractional polarizations are denoted P_D, for muons in diamagnetic environments, P_M for Mu, and P_R for radicals. For many systems the sum of observed polarizations is less than the initial polarization in the muon beam. A fraction P_L must have been lost at early times on the scale of the experiment. Because of this partial loss we have to distinguish between P_i and the actual fractional yield h_i of a chemical species. It is often not trivial to find the accurate relation between the two, but of course $P_i \leq h_i$ and $\sum h_i = 1$ holds.

Since it is difficult to judge a superposition of signals in time space it is common to transform the data to frequency space and display the Fourier power. For a quantitative analysis the parameters may be obtained in a direct fit of eqn. (2.4) to the time histogram. This involves four parameters per frequency. It is often cumbersome for multiline spectra. A fit in Fourier space allows a separate treatment of each frequency and is therefore often preferred [34]. A damped cosine (eqn. 2.5) transforms into a Lorentzian line with a line width parameter λ.

In general, there are two contributions to the relaxation parameter λ:

$$\lambda = \lambda_0 + \lambda_{ch}. \tag{2.6}$$

λ_{ch} is associated with the chemical reaction of interest. λ_0 is the relaxation of the signal in the absence of this reaction and may have to do with inhomogeneities of the magnetic field, with electron spin flips and other physical or even chemical relaxation processes.

Up to 10^8 muons are stopped in the sample in a typical μSR experiment. Each of them creates of the order of 10^6 radicals in its ionization track. In order to observe a bimolecular reaction of Mu or of muonated radicals with an added substrate S on a microsecond time scale the concentration $[S]$ must be $\gtrsim 10^{-5}$ M. Even if one assumes that all radicals in the track react with S the substrate will never deplete in the course of the experiment, i.e. one has

14

an ideal pseudo-first order kinetics,

$$\lambda_{ch} = k \cdot [S]. \tag{2.7}$$

For unimolecular reactions λ_{ch} is of course the rate constant directly.

Chapter 3

Theory

3.1   Hamiltonians, eigenvectors and energies

3.1.1   Muons in diamagnetic environments

Free muons, or muons in diamagnetic environments, are characterized by the Hamiltonian, in units of $\hbar$:

$$\hat{H} = -\omega_\mu \hat{I}_z^\mu, \tag{3.1}$$

where $\omega_\mu = 2\pi B \gamma_\mu = B \cdot 0.085161$ Mrad$\cdot$s$^{-1}\cdot$G^{-1} is the muon Larmor angular frequency and γ_μ is the is the muon gyromagnetic ratio. The transition observed in transverse fields at ω_μ is between the two Zeeman states, $m^\mu = \pm\frac{1}{2}$.

3.1.2   Muonium

Mu is a two-spin-$\frac{1}{2}$ system characterized by

$$\hat{H} = \omega_e \hat{S}_z - \omega_\mu \hat{I}_z^\mu + \omega_0 \hat{S}\cdot\hat{I}^\mu. \tag{3.2}$$

The muon and the electron are coupled by the isotropic hyperfine interaction, $\omega_0 = 2\pi \cdot A_\mu$. $\omega_e = 2\pi B \gamma_e = B \cdot 17.608$ Mrad$\cdot$s$^{-1}\cdot$G^{-1} is the electron Larmor angular frequency and γ_e its gyromagnetic ratio. The eigenvectors and eigenenergies are given in Table 3.1. In the absence of a magnetic field the four states combine to a triplet and a singlet system. The degeneracy of the triplet is lifted in a field B. The energies as a function of field are displayed in the form of a Breit-Rabi diagram in Figure 3.1. Four spectroscopic transitions are magnetic dipole allowed in low transverse fields (see Section 3.2.1 for a derivation of the selection rules). Two of them (broken

Table 3.1: Eigenvectors, energies and transverse field μSR transition frequencies for a muon-electron system with coupling constant ω_0.[†]

$$|\,1>=|\,\alpha^\mu\alpha^e> \qquad\qquad E^1 = \quad \tfrac{1}{4}\omega_0 + \omega_-$$
$$|\,2>=s\,|\,\alpha^\mu\beta^e> +c\,|\,\beta^\mu\alpha^e> \qquad E^2 = \quad \tfrac{1}{4}\omega_0 + \Omega$$
$$|\,3>=|\,\beta^\mu\beta^e> \qquad\qquad E^3 = \quad \tfrac{1}{4}\omega_0 - \omega_-$$
$$|\,4>=c\,|\,\alpha^\mu\beta^e> -s\,|\,\beta^\mu\alpha^e> \qquad E^4 = - \quad \tfrac{3}{4}\omega_0 - \Omega$$

$$\omega^{12} = E^1 - E^2 = \omega_- - \Omega \qquad \text{amplitude proportional} \quad \text{to } c^2$$
$$\omega^{23} = E^2 - E^3 = \omega_- + \Omega \qquad\qquad\qquad\qquad\qquad \text{to } s^2$$
$$\omega^{14} = E^1 - E^4 = \omega_- + \Omega + \omega_0 \qquad\qquad\qquad\qquad \text{to } s^2$$
$$\omega^{43} = E^4 - E^3 = \omega_- - \Omega - \omega_0 \qquad\qquad\qquad\qquad \text{to } c^2$$

[†] Eigenvectors in the basis of muon-electron Zeeman eigenstates, energies in units of $\hbar$

$$c = 2^{-\frac{1}{2}}\{1 + (\omega_e + \omega_\mu)/[\omega_0^2 + (\omega_e + \omega_\mu)^2]^{\frac{1}{2}}\}^{\frac{1}{2}}$$
$$s = 2^{-\frac{1}{2}}\{1 - (\omega_e + \omega_\mu)/[\omega_0^2 + (\omega_e + \omega_\mu)^2]^{\frac{1}{2}}\}^{\frac{1}{2}}$$
$$\omega_- = \tfrac{1}{2}(\omega_e - \omega_\mu)$$
$$\Omega = \tfrac{1}{2}\{[\omega_0^2 + (\omega_e + \omega_\mu)^2]^{\frac{1}{2}} - \omega_0\}$$

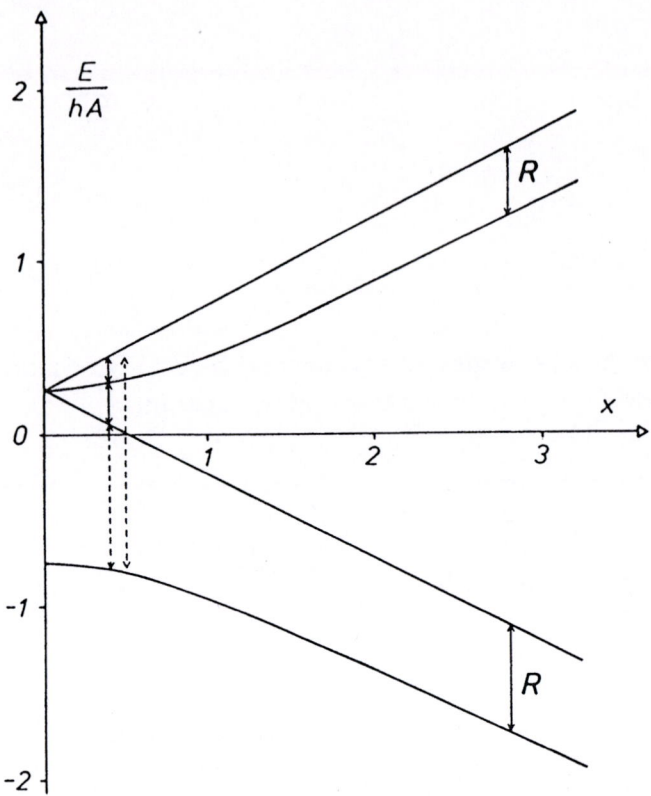

Figure 3.1: Breit-Rabi diagram for a two-spin-$\frac{1}{2}$ system. The magnetic field parameter x represents the ratio of Larmor to hyperfine frequency.

lines in Fig. 3.1) are of the order of A_μ, i.e. 4.5 GHz. They are resolved experimentally only with special efforts. It is therefore common to observe Mu by means of its triplet precessions, $\omega_{Mu} = 2\pi \cdot 1394$ kHz $\cdot$ G^{-1} in fields $\lesssim 20$ Gauss where the two transitions are degenerate. Mu is thus easily distinguished from muons in a diamagnetic environment, owing to the large magnetic moment of the electron.

In fields of a few kGauss or more only two transitions are allowed. They are designated R in Figure 3.1. They are purely nuclear, i.e. the analogues of ENDOR transitions for H.

3.1.3   Muonated free radicals

Most muonated organic radicals contain, apart from the muon, a considerable number of other magnetic nuclei k which are also coupled to the unpaired

---

$$|1>=|\alpha^\mu\alpha^e>\prod|K,M>$$
$$|2>=(s\,|\,\alpha^\mu\beta^e>+c\,|\,\beta^\mu\alpha^e>)\prod|K,M>$$
$$|3>=|\beta^\mu\beta^e>\prod|K,M>$$
$$|4>=(c\,|\,\alpha^\mu\beta^e>-s\,|\,\beta^\mu\alpha^e>)\prod|K,M>$$

$$E^1 = \tfrac{1}{4}\omega_0^\mu + \omega_- \quad -\sum\omega_k M + \tfrac{1}{2}\sum\omega_0^k M$$
$$E^2 = \tfrac{1}{4}\omega_0^\mu + \Omega \quad -\sum\omega_k M + \tfrac{1}{2}(c^2-s^2)\sum\omega_0^k M$$
$$E^3 = \tfrac{1}{4}\omega_0^\mu - \omega_- \quad -\sum\omega_k M - \tfrac{1}{2}\sum\omega_0^k M$$
$$E^4 = -\tfrac{3}{4}\omega_0^\mu - \Omega \quad -\sum\omega_k M - \tfrac{1}{2}(c^2-s^2)\sum\omega_0^k M$$

$$\omega^{12} = \omega_- - \Omega \quad +s^2\sum\omega_0^k M$$
$$\omega^{23} = \omega_- + \Omega \quad +c^2\sum\omega_0^k M$$
$$\omega^{14} = \omega_- + \Omega + \omega_0^\mu \quad +c^2\sum\omega_0^k M$$
$$\omega^{43} = \omega_- - \Omega - \omega_0^\mu \quad +s^2\sum\omega_0^k M$$

[†] Abbreviations as in Table 3.1 with $\omega_0 = \omega_0^\mu$.

electron. The corresponding spin Hamiltonian reads

$$\hat{H} = \omega_e\hat{S}_z - \omega_\mu\hat{I}_z^\mu + \omega_0^\mu\hat{S}\cdot\hat{I}^\mu - \sum_k\omega_k\hat{I}_z^k + \sum_k\omega_0^k\hat{S}\cdot\hat{I}^k. \qquad (3.3)$$

ω_0^μ and ω_0^k are the Fermi contact hyperfine interaction constants. In liquids, anisotropic contributions to the hyperfine coupling arising from direct dipole-dipole interaction are averaged to zero due to rapid tumbling of the radicals. The eigenfunctions, written in a basis of product functions,

$$|\chi_i>=|\chi_i^\mu>|\chi_i^e>\prod_k|\chi_i^k>, \qquad (3.4)$$

are given for a set of equivalent nuclei with total angular momentum operator $\hat{K} = \sum\hat{I}^k$ in Table 3.2, together with the first order energies and the μSR transition frequencies for transverse fields.

19

3.2 Evolution of spin polarization

We now turn to the theoretical description of the free induction decay of the muon spin polarization with particular attention to the effects of interconversion of chemical species and concomitant loss of polarization. Several authors have worked out proceedures for such an analysis [35-38]. Here, we give a short outline of a method based on a state density approach which allows us to discuss the results of experiments with muonated radicals [26,39].

The observable muon polarization along a direction u is the sum of contributions $P_u^X(t)$ from all species X containing muons,

$$P_u(t) = \sum_X P_u^X(t). \tag{3.5}$$

We consider the general case of a species which has been formed at time t' $(0 \leq t' \leq t)$ and still exists at time t. Its contribution is given by

$$P_u^X(t) = \int_0^t p_u^X(t') q^X(t,t') k_0 dt', \tag{3.6}$$

where $q^X(t,t') k_0$ is the probability per unit time that X has been formed at time t' with a pseudo-first order rate constant k_0 *and* that it still exists at time t. $p_u^X(t)$ is the polarization of that fraction of the muon ensemble defined by $q^X(t,t')$. It is computed in the Heisenberg representation as the expectation value of the u-component of the Pauli spin operator, $\hat{\sigma}_u^X$,

$$p_u^X(t) = < \hat{\sigma}_u^X(t) > = Tr\{\hat{\rho}(0) \cdot \hat{\sigma}_u^X(t)\}. \tag{3.7}$$

$\hat{\sigma}_u^X$ is a matrix of dimension $N = 4\Pi_k(2I^k + 1)$. It is expressed most easily in a basis corresponding to quantization along the axis of observation

$$\hat{\sigma}_u^X(0) = \hat{\sigma}_u^\mu(0) \otimes \mathbf{1}^e \otimes \Pi_k \mathbf{1}^k = \begin{pmatrix} 1 & 0 \\ 0 & -1 \end{pmatrix} \otimes \mathbf{1}^e \otimes \Pi_k \mathbf{1}^k. \tag{3.8}$$

For $0 \leq t \leq t'$ it evolves in time according to the Hamiltonian $\hat{H}'$ of the precursor

$$\hat{\sigma}_u^X(t') = e^{i\hat{H}'t'/\hbar} \hat{\sigma}_u^X(0) e^{-i\hat{H}'t'/\hbar}, \tag{3.9}$$

and for $t > t'$ according to $\hat{H}$ of the observed species X

$$\hat{\sigma}_u^X(t) = e^{i\hat{H}(t-t')/\hbar} \hat{\sigma}_u^X(t') e^{-i\hat{H}(t-t')/\hbar}. \tag{3.10}$$

$\hat{\rho}(0)$ is given conveniently in a basis corresponding to quantization of the spins along the direction b of beam polarization. For 100% initial polarization it is

$$\begin{aligned}\hat{\rho}(0) &= \hat{\rho}^{\mu}(0) \otimes \hat{\rho}^{e}(0) \otimes \Pi_k \hat{\rho}^k(0) \\ &= \frac{1}{2}\begin{pmatrix} 2 & 0 \\ 0 & 0 \end{pmatrix} \otimes \frac{1}{2} \cdot \mathbf{1}^e \otimes \frac{4}{N}\Pi_k \mathbf{1}^k \\ &= \frac{1}{N}(\mathbf{1}^{\mu} + \hat{\sigma}_b) \otimes \mathbf{1}^e \otimes \Pi_k \mathbf{1}^k. \end{aligned} \qquad (3.11)$$

Inserting (3.11) into (3.7) and noting that $Tr\{\hat{\sigma}_u\} = 0$, one obtains

$$p_u^X(t) = N^{-1} Tr\{\hat{\sigma}_b \cdot \hat{\sigma}_u^X(t)\}. \qquad (3.12)$$

This is to be used in the integration (3.6) for the different kinetic cases.

3.2.1 Evaluation for a general species formed at time zero

We assume that the observed species X is formed at $t' = 0$ with probability h_X, and that it decays with a pseudo-first order rate constant k_1. Thus

$$q^X(t, t') \cdot k_0 = h_X \cdot e^{-k_1 t} \cdot \delta(t' - 0), \qquad (3.13)$$

and the integration (3.6) is performed readily. Usually, $P_u^X(t)$ has to be evaluated only for the special cases $b = u = z$ (longitudinal field) or $b = u = x$ (transverse field), i.e.

$$\begin{aligned} P_u^X(t) &= N^{-1} \cdot h_X \cdot e^{-k_1 t} \sum_{m,n} |<m\,|\,\hat{\sigma}_u\,|\,n>|^2\, e^{i\omega^{nm}t} \\ &= N^{-1} \cdot h_X \cdot e^{-k_1 t} \sum_{m} \Big\{ |<m\,|\,\hat{\sigma}_u\,|\,m>|^2 \\ &\quad + 2 \sum_{n<m} |<m\,|\,\hat{\sigma}_u\,|\,n>|^2 \cos(\omega^{nm}t) \Big\}, \end{aligned} \qquad (3.14)$$

where $|\,m>$, $|\,n>$ are the eigenkets of $\hat{H}$ with energies $\hbar\omega^m$ and $\hbar\omega^n$, respectively. The magnitude of the muon polarization oscillating at a transition frequency ω^{nm} is proportional to the transition moment of the muon spin operator. This reveals the close relationship between μSR and magnetic resonance. By expressing the eigenkets in terms of the basis (3.4)

$$|\,m> = \sum_i c_{im}\,|\,\chi_i>, \qquad (3.15)$$

21

eq. (3.14) becomes for $u = z$

$$P_z^X(t) = \frac{hX}{N} e^{-k_1 t} \sum_m \left\{ | \sum_i c_{im}^* c_{im} 2m_i^\mu |^2 \right.$$

$$\left. + 2 \sum_{n<m} | \sum_i c_{im}^* c_{in} 2m_i^\mu |^2 \cos(\omega^{nm} t) \right\}, \qquad (3.16)$$

and for $u = x$

$$P_x^X(t) = \frac{2hX}{N} e^{-k_1 t} \sum_m \sum_{n<m} | \sum_{i,j} c_{im}^* c_{jn} < \chi_i^\mu | \hat{\sigma}_x^\mu | \chi_j^\mu >$$

$$\times \delta_{ij}^e \Pi_k \delta_{ij}^k |^2 \cos(\omega^{nm} t). \qquad (3.17)$$

Selection rules

For P_z^X there are only contributions from transitions between states $| m >$ and $| n >$ which contain identical basis kets $| \chi_i >$, whereas for P_x^X there are contributions between states $| m >$ and $| n >$ from basis kets $| \chi_i >$ and $| \chi_j >$ differing in the muon but not in the electron and nuclear parts. Since a Hamiltonian (3.2, 3.3) mixes only product kets with equal total magnetic quantum number $M = m^\mu + m^e + \sum_k m^k$ this leads immediately to the general selection rules

$$\begin{aligned} \Delta M &= 0 \quad \text{for} \quad u = z, \\ \Delta M &= \pm 1 \quad \text{for} \quad u = x. \end{aligned} \qquad (3.18)$$

More specific selection rules have been derived for low and zero fields, and expressions to calculate the total number of transitions for muonated radicals and related species have been given [26,39]. Furthermore, numerical examples were presented to illustrate the appearence of radical spectra for these conditions. Experimental observation is possible only in selected cases [26,40].

In the high field limit the eigenkets $| m >$ and $| n >$ become equal to individual product kets $| \chi_i >$ since the spins are effectively decoupled. Thus, the eigenstates are characterized by m^μ, m^e and all m^k, and the coefficients in (3.15-3.17) are usually equal to zero or one. For this case (3.16) leads to

$$P_z^X = hX e^{-k_1 t}, \qquad (3.19)$$

i.e. for $k_1 = 0$ there is no evolution of spin polarization in a strong longitudinal field, except under conditions of *avoided level crossings* (vide infra). Eqn. (3.18) yields the selection rules

$$\Delta m^\mu = \pm 1, \quad \Delta m^e = \Delta m^k = 0 \quad \text{for} \quad u = x. \qquad (3.20)$$

Simplifications for radicals in high fields

In high fields ($\omega_e \gg \omega_0^\mu, \omega_0^k$), perturbation theory is applied. The simplest approach treats all hyperfine interactions as perturbations. This leads to the first-order energies, in units of $\hbar$:

$$E^i = \omega_e m_i^e - \omega_\mu m_i^\mu - \sum_k \omega_k m_i^k + \omega_0^\mu m_i^e m_i^\mu + \sum_k \omega_0^k m_i^e m_i^k. \qquad (3.21)$$

With the selection rules (3.20) one derives immediately that all transitions degenerate to two only. They correspond to those denoted R in Figure 3.1. The frequencies obey the ENDOR condition and appear at

$$|\nu| = |\nu_\mu \pm \frac{1}{2} \cdot A_\mu|. \qquad (3.22)$$

Since for most cases the muon hyperfine coupling is larger than the coupling of the other nuclei the following treatment gives more accurate results: The muon-electron Hamiltonian (3.2) is taken as $\hat{H}^0$, and only the nuclear terms are treated as perturbations, i.e. (3.3) is rewritten as

$$\hat{H} = \hat{H}^0 + \sum_k \omega_k \hat{I}_z^k + \sum_k \omega_0^k \hat{S} \cdot \hat{I}^k. \qquad (3.23)$$

The results are given in first order in Table 3.2. As for Mu, the amplitudes of ω^{23} and ω^{14} are proportional to s^2, i.e. they vanish in high fields. $|A_\mu|$ is determined accurately from the difference (for $|A_\mu| \leq |2\nu_\mu|$) or in most cases (where $|A_\mu| \geq |2\nu_\mu|$) from the sum of the two frequencies.

As the magnetic field decreases the effect of the additional magnetic nuclei results in a broadening and then in a splitting of the lines. In first order, the splitting is $s^2 A_k$ for each nucleus, and a multiplet pattern which is typical for magnetic resonance is obtained. In particular, a single proton leads to a doublet. This effect is used to aid the assignment of radicals.

The conclusion is that the nuclei k may be completely neglected in decoupling fields. This is particularly convenient for the evaluation of expressions (3.13), as the dimension of the matrices reduces to four. Exceptions occur for the case of accidental degeneracies of states, which renders first-order treatment inappropriate. This is investigated next.

The case of avoided level crossings (ALCs)

In atomic spectroscopy and nuclear quadrupole resonance the mixing effect of near degenerate levels has been known for many years [41,42]. It was *Abragam*[27] who proposed the use of the phenomenon to detect nuclear

23

quadrupole splittings, tunnelling energy separations in molecular crystals or hyperfine splittings of paramagnetic ions in longitudinal field μSR. For muonated free radicals oscillations in the time differential spectrum $P_z^X(t)$ are predicted which lead to resonances in the time integrated quenching curve $\overline{P_z^X}(B) = A$ (eqn. 2.3). A theoretical treatment developped for high fields by *Heming et al.*[29] starts with the general expression for $P_z^X(t)$ (eqn. 3.16). Consider first a three spin system with muon, electron and one nucleus. From eqn. (3.21) we see immediately that states related by $\Delta(m^\mu + m^k) = 0$, e.g. $|\alpha^e>|\beta^\mu>|m^k>$ and $|\alpha^e>|\alpha^\mu>|m^k - 1>$ cross in energy at a resonant field

$$B_r = \frac{|\omega_0^\mu - \omega_0^k|}{2(\omega^\mu - \omega^k)} = \frac{|A_\mu - A_k|}{2(\gamma_\mu - \gamma_k)}. \qquad (3.24)$$

For $A_\mu > A_k$ only the states with $m^s = \frac{1}{2}$ cross, for $A_\mu < A_k$ those with $m^s = -\frac{1}{2}$. In higher order crossing is avoided by coupling of the two states to energetically distant states via non-secular terms of the Hamiltonian. One obtains

$$B_r(M) = \left| \frac{A_\mu - A_k}{2(\gamma_\mu - \gamma_k)} - \frac{A_\mu^2 - 2MA_k^2}{2\gamma_e(A_\mu - A_k)} \right|, \qquad (3.25)$$

and the energy gap or ALC transition frequency becomes

$$\omega(M,B) = \pi \left[\left(2B(\gamma_\mu - \gamma_k) - |A_\mu - A_k| + \frac{A_\mu^2 - 2MA_k^2}{2B\gamma_e} \right)^2 \right.$$

$$\left. + \left(\frac{cA_\mu A_k}{B\gamma_e} \right)^2 \right]^{1/2}, \qquad (3.26)$$

and

$$\omega_r(M, B_r) \approx \frac{\pi c A_\mu A_k}{B_r \gamma_e}, \qquad (3.27)$$

where $I^k \geq M \geq (-I^k + 1)$ and $c = [I^k(I^k + 1) - M(M - 1)]^{1/2}$. We thus expect $2I_k$ nearly degenerate ALC transitions. The coefficients of the mixing states are calculated from the secular equation and inserted into eqn. (3.16). Integration over time and normalization (eqn. 2.3) leads to [29]

$$\overline{P_z^X}(B) = h_X - \frac{(2h_X/N)\sum_{i=1}^{2I^k} \omega_{ri}^2}{k_1^2 + \omega_{ri}^2 + [2\pi(B - B_r)(\gamma_\mu - \gamma_k)]^2}, \qquad (3.28)$$

which is a sum of Lorentzian lines with full width at half maximum

$$\Delta B_{1/2}(M) = \frac{\omega_r}{\pi(\gamma_\mu - \gamma_k)}(1 + k_1^2/\omega_r^2)^{1/2}$$

$$= \frac{2cA_\mu A_k}{\gamma_e |A_\mu - A_k|}(1 + k_1^2/\omega_r^2)^{1/2} \qquad (3.29)$$

24

and amplitude

$$\overline{P_z^X}(B_r) = h_X - \frac{(2h_X/N)\omega_{ri}^2}{k_1^2 + \omega_{ri}^2}. \tag{3.30}$$

k_1 describes any process which takes the muon out of the ALC resonance, i.e. the muon decay and usually chemical reactions. T_1^e processes, however, require a more sophisticated treatment since they bring new muons into resonance, which is not taken into account in eqn. (3.13). A treatment for spin exchange processes has been given elsewhere [43].

The formula for $\Delta B_{1/2}$ shows that the resonances are narrow compared with the field range over which they are distributed. This allows the separate treatment of inequivalent nuclei. Equivalent nuclei are combined using the total nuclear spin operator $\hat{K} = \sum \hat{I}^k$ as in Table 3.2. This reduces the problem to a number of three-spin systems with degeneracies $g(K)$.

The major advances of the ALC technique over conventional transverse field μSR are the following:

- The coupling constants of nuclei other than the muon are obtained with the same precision as in ENDOR spectroscopy. This allows a complete characterization of the system and a safer assignment of a structure.

- Radicals will be observed even when a Mu precursor has a lifetime of a microsecond, as long as the transition frequency ω_r is high enough to produce a significant ALC signal in the 'remaining' muon lifetime. The loss of polarization during the precursor stage is negligible in high longitudinal fields [43]. In contrast, in transverse field experiments the muon polarization is lost when the Mu precursor lifetime exceeds $\approx 10^{-10}$s (see eqns. 3.33, 3.34).

- The constraint of time differential μSR to have no more than one muon at a time in the sample is lifted. This allows taking advantage of the high muon fluxes which become available at modern accelerators.

3.2.2 Evaluation for a slowly formed radical

We now consider the case of radical formation ($X = R$) via two parallel channels according to the scheme

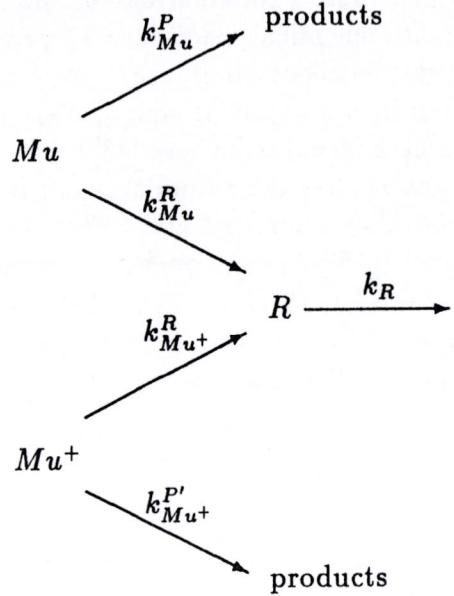

under conditions where there is non-negligible evolution of the muon spin in the precursors, which are here assumed to be Mu and Mu^+. The muon polarization in R is the vector sum of the contributions from the two channels. We first analize the fraction transferred from Mu. For this we have

$$q^R(t, t')k_{Mu}^R = h_{Mu}k_{Mu}^R e^{-t'/\tau_{Mu}} \cdot e^{-k_R(t-t')}, \qquad (3.31)$$

with the precursor lifetime $\tau_{Mu} = (k_{Mu}^R + k_{Mu}^P)^{-1}$. The polarization in a stable radical ($k_R \approx 0$) at times long compared with τ_{Mu} but short compared with the inverse frequency to be observed is obtained by integration:

$$
\begin{aligned}
P_u^R(t) = \ &\frac{1}{4}h_{Mu}\tau_{Mu}k_{Mu}^R \cdot \sum_{k,l,m,n} \left\{ (1 + \Delta_{lknm}^2)^{-1} \right. \\
&\times <m\,|\,\hat{\sigma}_u\,|\,n><n\,|\,l><l\,|\,\hat{\sigma}_u\,|\,k><k\,|\,m> \\
&\times \left. [\cos(\omega_R^{nm}t) - \Delta_{lknm} \cdot \sin(\omega_R^{nm}t)] \right\},
\end{aligned}
\qquad (3.32)
$$

with $\Delta_{lknm} = \tau_{Mu}(\omega_{Mu}^{lk} - \omega_R^{nm})$. $|\,k>$ and $|\,l>$ denote the eigenstates of Mu, and $|\,n>$ and $|\,m>$ those of the radical. We are interested in the solution for transverse fields which are high for the radical ($\omega_e \gg \omega_0^\mu, \omega_0^k$, i.e. $s_R^2 \approx 0$, $c_R^2 \approx 1$) but not necessarily for Mu. These are the conditions where only

26

two radical frequencies, ω_R^{12} and ω_R^{43}, have to be considered, and we can work with matrices of rank 4. The matrix elements in (3.32) restrict polarization transfer to ω_R^{12} from the precursor frequencies ω_{Mu}^{12} and ω_{Mu}^{14}, and to ω_R^{43} from ω_{Mu}^{43} and ω_{Mu}^{23}. This is indicated with arrows in the theoretical spectrum in Figure 3.2. The explicit results for the x-components at $t = 0$ are

$$P_x^{12}(0) = \frac{1}{2} h_{Mu} \tau_{Mu} k_{Mu}^R \left\{ \frac{c_{Mu}^2}{1 + \Delta_{1212}^2} + \frac{s_{Mu}^2}{1 + \Delta_{1412}^2} \right\},$$

$$P_x^{43}(0) = \frac{1}{2} h_{Mu} \tau_{Mu} k_{Mu}^R \left\{ \frac{c_{Mu}^2}{1 + \Delta_{4343}^2} + \frac{s_{Mu}^2}{1 + \Delta_{2343}^2} \right\}, \qquad (3.33)$$

and similarly for the y-components

$$P_y^{12}(0) = \frac{1}{2} h_{Mu} \tau_{Mu} k_{Mu}^R \left\{ \frac{c_{Mu}^2 \cdot \Delta_{1212}}{1 + \Delta_{1212}^2} + \frac{s_{Mu}^2 \cdot \Delta_{1412}}{1 + \Delta_{1412}^2} \right\},$$

$$P_y^{43}(0) = \frac{1}{2} h_{Mu} \tau_{Mu} k_{Mu}^R \left\{ \frac{c_{Mu}^2 \cdot \Delta_{4343}}{1 + \Delta_{4343}^2} + \frac{s_{Mu}^2 \cdot \Delta_{2343}}{1 + \Delta_{2343}^2} \right\}. \qquad (3.34)$$

Δ is the important parameter in eqns. (3.33, 3.34). It is a function of the frequency difference between precursor and product and depends therefore on the external magnetic field. Furthermore, if k_{Mu}^P describes a scavenging reaction it can be influenced by variation of the scavenger concentration. In particular, for $\Delta \ll 1$, $P_y^{nm}(0)$ vanishes, i.e. the muon spins in the radical initially point in the same direction as originally in the beam. If in addition $k_{Mu}^P = 0$ we have full transfer of polarization into the product, $P_x^{nm}(0) = | P^{nm} | = 0.5 h_{Mu}$, otherwise a fraction $k_{Mu}^P/(k_{Mu}^R + k_{Mu}^P)$ is transferred to the products of the scavenging reaction.

Similar equations are readily obtained for the second channel, the only difference being in the Hamiltonian of the precursor. Mu^+ can be regarded as Mu in the limit of zero for the hyperfine coupling. This leads to the expected one-frequency spectrum with $\omega_D = -\omega_\mu$ (see Table 3.1 with $c = 1$ and $s = 0$). The solutions for the components of the muon polarization in the radical are also contained in eqns. (3.33, 3.34), again for $c_{Mu} = 1$ and $s_{Mu} = 0$, and for $\Delta_{nm} = \tau_D(\omega_D - \omega_R^{nm})$, where $\tau_D = (k_{Mu^+}^{P'} + k_{Mu^+}^R)^{-1}$ is the lifetime of the diamagnetic precursor. It is important to recognize that Δ and therefore P_y^{nm} is of opposite sign for Mu^+ instead of Mu as precursor because of $| \omega_D | < | \omega_R^{nm} | < | \omega_{Mu}^{lk} |$, which is also seen in Figure 3.2.

The observed magnitude of the residual polarization on a radical frequency is thus given by

$$| P^{nm} | = \left\{ \left[\sum P_x^{nm}(0) \right]^2 + \left[\sum P_y^{nm}(0) \right]^2 \right\}^{1/2}, \qquad (3.35)$$

27

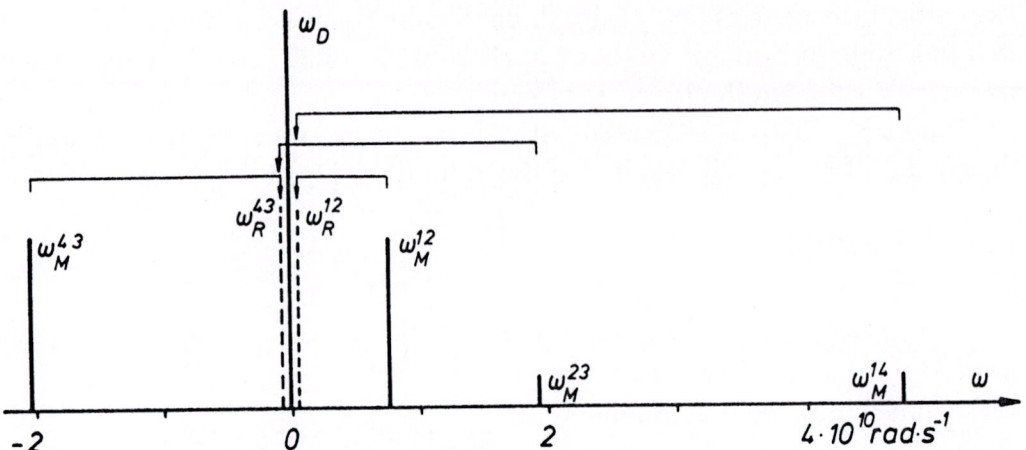

Figure 3.2: Theoretical μSR spectrum for equal amounts of fully polarized muons in diamagnetic environments, in a muonated radical with $\omega_0^\mu = 2\pi \cdot 200$ MHz, and in Mu, for a transverse magnetic field of 1600 Gauss.

and the corresponding residual phase

$$\Phi^{nm} = \arctan \frac{\sum P_y^{nm}(0)}{\sum P_x^{nm}(0)}, \tag{3.36}$$

where the sum goes over the two channels. Because of the different signs of Δ the contributions to $P_y^{nm}(0)$ might accidentally cancel and cause small residual phases despite non-negligible precursor lifetimes. Such a situation has been observed [44].

3.2.3 Evaluation for a slowly formed diamagnetic species

The treatment of the formation of a diamagnetic species according to

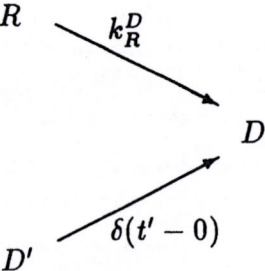

can be regarded as a special case of slow radical formation (Section 3.2.2) in the limit of zero hyperfine coupling for the product and a reduced coupling for the paramagnetic precursor. We obtain for the components of the polarization from eqns. (3.33, 3.34) for moderately high fields ($s_R^2 \cong 0$)

$$
\begin{aligned}
P_x^D(0) &= \frac{1}{2} h_R \tau_R k_R^D \left\{ \frac{1}{1 + \Delta_{12D}^2} + \frac{1}{1 + \Delta_{43D}^2} \right\}, \\
P_y^D(0) &= \frac{1}{2} h_R \tau_R k_R^D \left\{ \frac{\Delta_{12D}}{1 + \Delta_{12D}^2} + \frac{\Delta_{12D}}{1 + \Delta_{43D}^2} \right\}.
\end{aligned}
\qquad (3.37)
$$

The second channel leaves no time for muon precession in the precursor and gives a contribution $P_x^D(0) = h_{D'}$.

In the high field limit the radical lines are displaced symmetrically about ω_D (eqn. 3.22). Hence we have $\Delta_{12D} = -\Delta_{43D}$, and $P_y^D(0)$ and the residual phase vanishes.

Chapter 4

The cyclohexadienyl radical

4.1 Observation in the liquid phase

4.1.1 Transverse field μSR

The cyclohexadienyl radical, C_6H_7, was often observed by ESR of irradiated benzene [45,23]. Similarly, C_6D_6H was detected upon X-irradiation of benzene-d_6 in adamantane [46]. The muonated analogues, C_6H_6Mu and C_6D_6Mu, were among the first organic radicals detected by μSR [8,47]. Corresponding FT-μSR spectra are shown for different magnetic fields in Figure 4.1. Only the radical lines are displayed. As in all μSR spectra of organic liquids, there is a further strong line at a low frequency corresponding to ω_D for muons in diamagnetic environments. As a function of field the radical lines behave as described in Table 3.2 and on page 23. They are placed symmetrically around $\frac{1}{2}A_\mu$ (broken line). In high fields, here 3000 G, there are two lines. The sum of the two frequencies gives the muon-electron hyperfine coupling constant directly. We obtain $A_\mu = 514.6$ MHz for C_6H_6Mu and $A_\mu = 520.0$ MHz for C_6D_6Mu, which corresponds to *reduced coupling constants*

$$A'_\mu = A_\mu \cdot \mu_p / \mu_\mu = 0.3142 \cdot A_\mu \qquad (4.1)$$

of 161.7 MHz and 163.4 MHz, respectively. At 1000 G the lines of C_6H_6Mu split into doublets by 1.23 MHz, which is slightly less than in a less accurate early determination [26]. The splitting increases to 2.5 MHz at 700 G and 4.9 MHz at 500 G. Based on perturbation treatment (page 23) this corresponds to about 150 MHz for the coupling constant of a proton. Accurate numerical simulation gives 125 MHz. At 500 G the lines are broad and show a tendency to further splitting which is due to nuclei with smaller couplings. For C_6D_6Mu the lines are still unsplit at 1000 G and broad with an indication of splitting

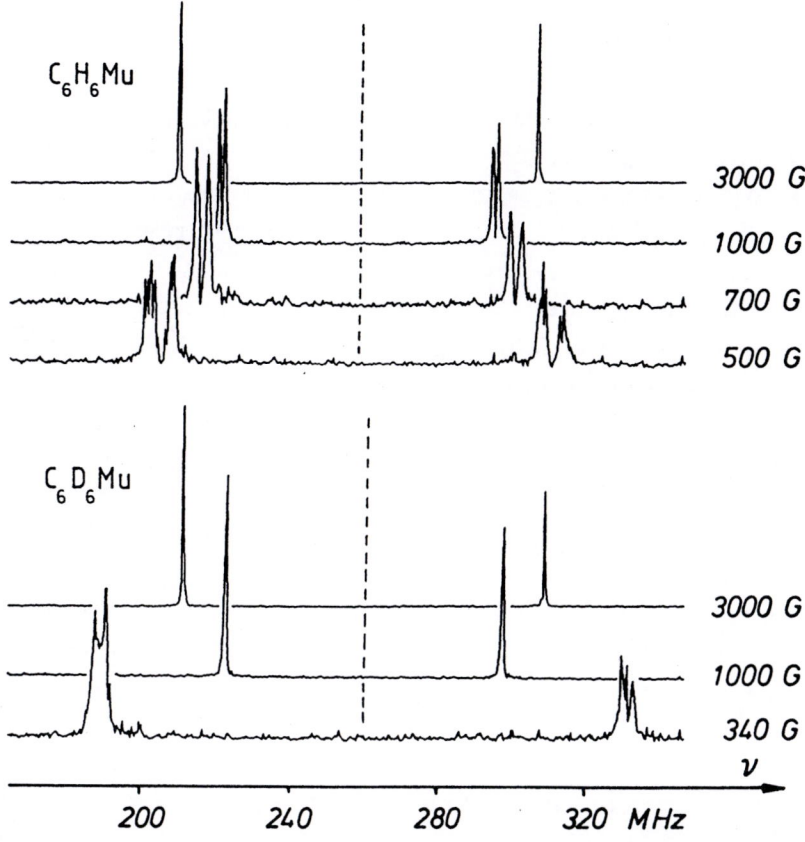

Figure 4.1: Fourier power μSR spectra obtained with benzene.

at 340 G. The behaviour is understood in terms of the radical structure. Based on the ESR data one would have expected values of $A'_\mu = A_p = 134.1$ MHz for the methylene nuclei, in the absence of isotope effects. A deuteron in place of a proton in the methylene position does not lead to splitting in a field of 1000 G since its magnetic moment is a factor of 6.51 smaller than that of the proton. This effect will be used to aid the assignment in substituted cyclohexadienyl radicals.

4.1.2 Avoided level crossing μSR

According to the theory (eqn. 3.25) one expects to observe four ALC resonances for the unsubstituted cyclohexadienyl radical, one each for the *ortho*, *meta*, *para* and the methylene protons. Figure 4.2 shows a simulation using a muon hyperfine coupling constant A'_μ=161.7 MHz and a methylene proton coupling A_p=125 MHz as obtained from the transverse field experiments.

31

For the *ortho*, *meta*, and *para* protons A_p=-25.5 MHz, 7.5 MHz, and -36.8 MHz, respectively, as inferred from [46], were used. For $A_\mu > 0$, the lines to the right of the dividing vertical line correspond to negative proton couplings, the others to positive ones. The relative signs are not usually obtained from conventional magnetic resonance type experiments, while ALC yields them without special efforts. Unlike other magnetic resonance techniques, the signal amplitudes are not heavily dependent on the number of nuclei in resonance. High amplitudes require that the resonance frequency ω_r competes with the muon decay rate (eqn. 3.30). It means that for a given radical high resonance fields and low nuclear couplings are unfavourable (eqn. 3.27). This is demonstrated nicely in the simulation, where the low *meta* coupling leads to a very small signal. The effect of the greater magnitude of the *para* compared with the *ortho* coupling is compensated by the higher resonance field, so that the two corresponding signal amplitudes are the same.

The experimental spectrum was first reported by *Percival et al.* [48] for both C_6H_6Mu and C_6D_6Mu. Figure 4.3 displays our own results. Only three resonances were observed. Their shape conforms with the theoretically predicted Lorentzian line shape function, and their position is close to expectation. The resulting coupling constants are given in Table 4.1, along with Percival's values and with literature data for the hydrogen analogoues. Within error, there is no isotope effect for the ring protons, but considerable deviations are found for the methylene nuclei. They will be subject to discussion in Section 4.3.

For demonstration purpose we also show the time resolved longitudinal field measurement for the 2.08 T line. Top of Figure 4.4 displays the histogram obtained with pure benzene exactly on-resonance. The oscillation corresponds directly to the energy gap between the mixing states. The frequency of 0.60 MHz agrees with expectation (eqn. 3.27), and it is gratifying that the relaxation of 0.22 μs^{-1} is less than what is found in transverse fields (compare intersepts in Figure 8.2) where additional relaxation processes may operate. Each field point in an ALC spectrum is the time integral of such a histogram. The second picture is for a $2 \cdot 10^{-4}$ M solution of DPPH in benzene. Reaction of C_6H_6Mu with DPPH (see Section 8.3 for details) shows up as a strong relaxation. The bottom picture is for pure benzene again, but 50 mT off-resonance. It demonstrates stationary muon polarization at this field.

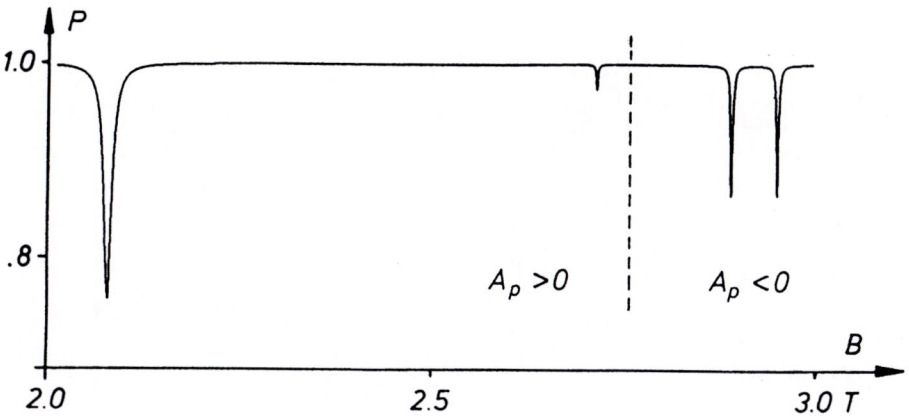

Figure 4.2: Simulated ALC spectrum of C_6H_6Mu.

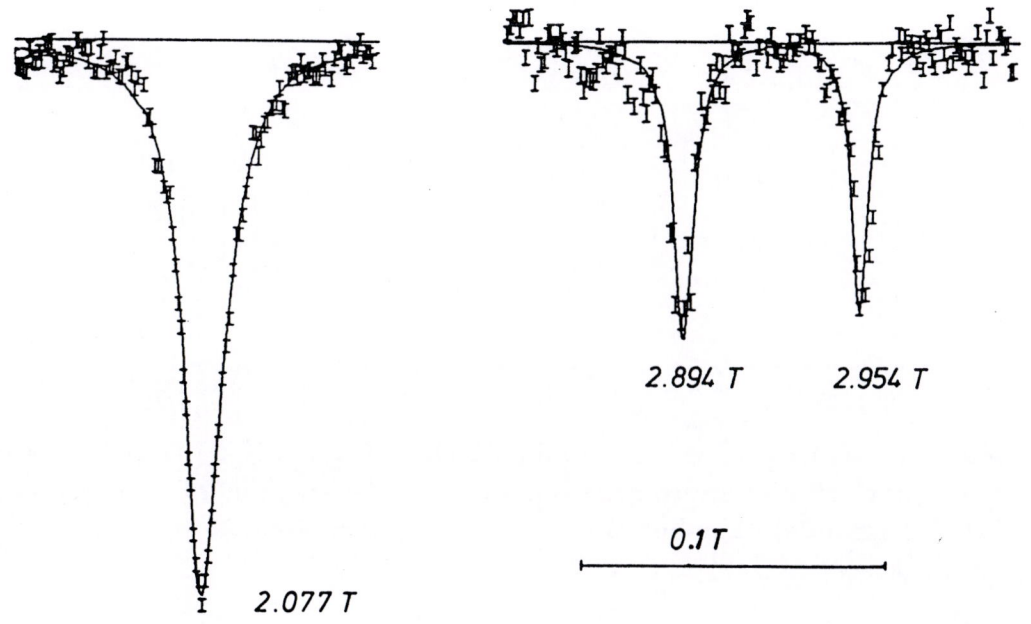

Figure 4.3: ALC resonances obtained with benzene.

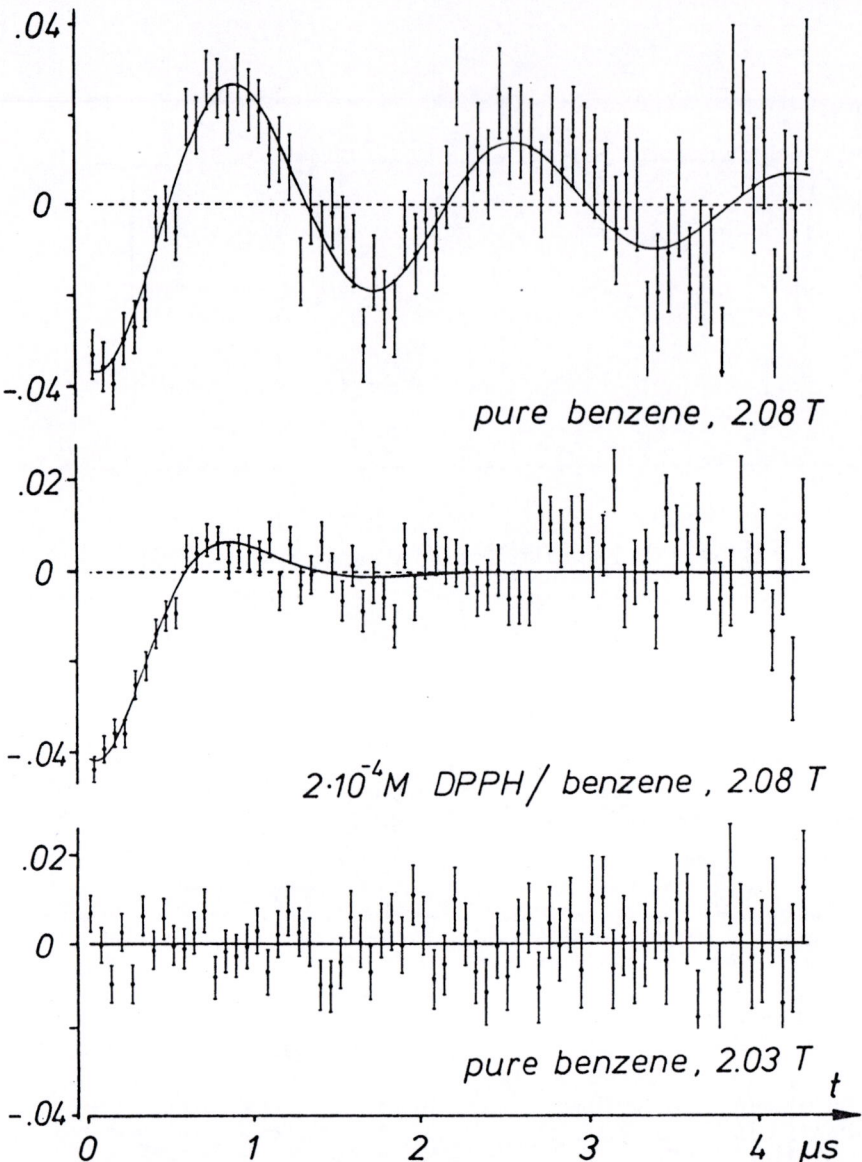

Figure 4.4: Time resolved longitudinal field histograms for the 2.08 T line on-resonance obtained with pure benzene (top) and with $2 \cdot 10^{-4}$ M solution of DPPH (middle). The bottom spectrum was measured off-resonance with pure benzene.

Table 4.1: reduced hyperfine coupling constants in MHz for isotopic cyclohexadienyl radicals

radical	methylene		ortho	meta	para	ref.
						nucleus
C_6H_6Mu	161.6 (Mu)	126.1 (H)	-25.1	+7.5	-36.2	[48]
C_6H_6Mu	161.7 (Mu)	126.2 (H)	-25.2	unobs.	-36.3	a)
C_6D_6Mu	163.3 (Mu)	124.3 (D)	-25.5	unobs.	-36.4	[48]
C_6H_7	134.1 (H)	134.1 (H)	25.3	7.8	37.1	[46]
C_6D_6H	136.2 (H)	144.0 (D)	-	-	-	[46]

a) This work

4.2 Observation in other phases

4.2.1 The first experiment with a single crystal

The cyclohexadienyl radical was detected in polycrystalline benzene at 232 K [49]. The reorientational motion is frozen or highly hindered, and anisotropic contributions to the hyperfine interaction are no longer averaged to zero. Therefore, the lines are inhomogeneously broadened.

The first and hitherto only observation of a muonated organic radical in a crystal was a substituted cyclohexadienyl radical in durene [50]. A 7.5 g single crystal was grown from a saturated solution in diethyl ether by slow evaporation of the solvent over a period of two months in a temperature regulated room. Durene forms a monoclinic crystal of space group $P2_1/a$ with two molecules per unit cell. For the experiments it was rotated around each of the orthogonal axes a, b, c' in intervals of 30°. A selection of spectra is given in Figure 4.5. For general orientations we observe eight radical lines. For $b \perp B_0$ and for B_0 parallel to one of the crystallographic axes they degenerate to four lines. From the analysis of the orientation dependence of the coupling constants we obtain the hyperfine principal values and anisotropies which are given in Table 4.2.

The two durene molecules per unit cell are symmetry related by a glide plane perpendicular to b. Inspection of the hyperfine couplings and of the direction cosines [50] shows that this holds also for the radicals. We find the chemically inequivalent radicals I and II which are symmetry related to their duplicates I' and II'.

In principle Mu can add to durene at the $C - H$ or at the $C - CH_3$ positions. Since the molecular centre is the only molecular symmetry element which is

35

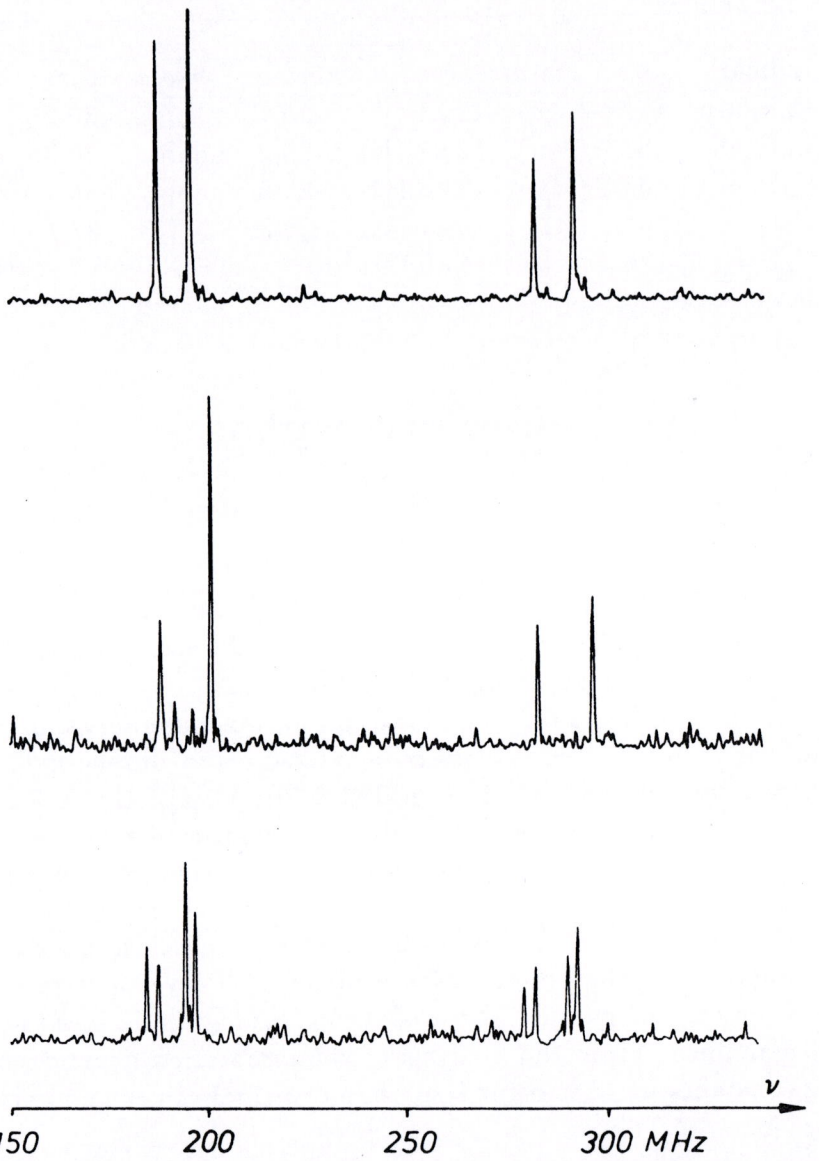

Figure 4.5: Radical frequencies observed in a durene single crystal.
Top: $b \| B_0$, middle: $b \perp B_0, \angle(c', B_0) = 150°$, bottom: $c' \perp B_0, \angle(a, B_0) = 30°$.

Table 4.2: Hyperfine principal values and anisotropies of the radicals observed in durene[a)]

radical	A'_{iso}	B'_{XX}	B'_{YY}	B'_{ZZ}
I	151.8	-2.9	0.4	2.6
I'	151.8	-2.6	0.3	2.3
II	145.6	-2.9	0.2	2.7
II'	145.6	-2.9	0.3	2.7

[a)]Values in MHz, $A'_{ii} = A'_{iso} + B'_{ii}$

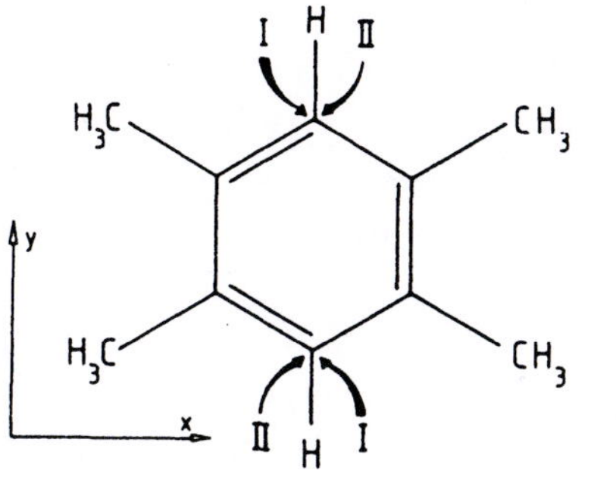

Figure 4.6: Sites for Mu addition to durene.

in the latter case. The fact that only two are observed suggests addition at the C−H sites from above and below the molecular plane (Figure 4.6). It is usually found that radicals produced in organic crystals by ionizing radiation are oriented nearly the same as the parent molecules [51]. Comparison of the orientation of the radical hyperfine tensor with the molecular orientation shows only a few degrees deviation and confirms Mu addition at the C−H site [50]. The values of A'_{iso} are lower than those found for the radical in benzene. This is a consequence of methyl substitution (see Chapter 5). The difference between I and II may be due to small distortions depending on the different crystallographic environments of Mu. Owing to the large distance between the muon and the delocalized unpaired electron one finds only a small anisotropy.

4.2.2 Observation of a surface adsorbed radical

NMR spectroscopy has become a major technique to characterize the interaction and mobility of organic molecules adsorbed on surfaces [52,53]. Little is known in this field for free radicals. Recently, it became possible to study

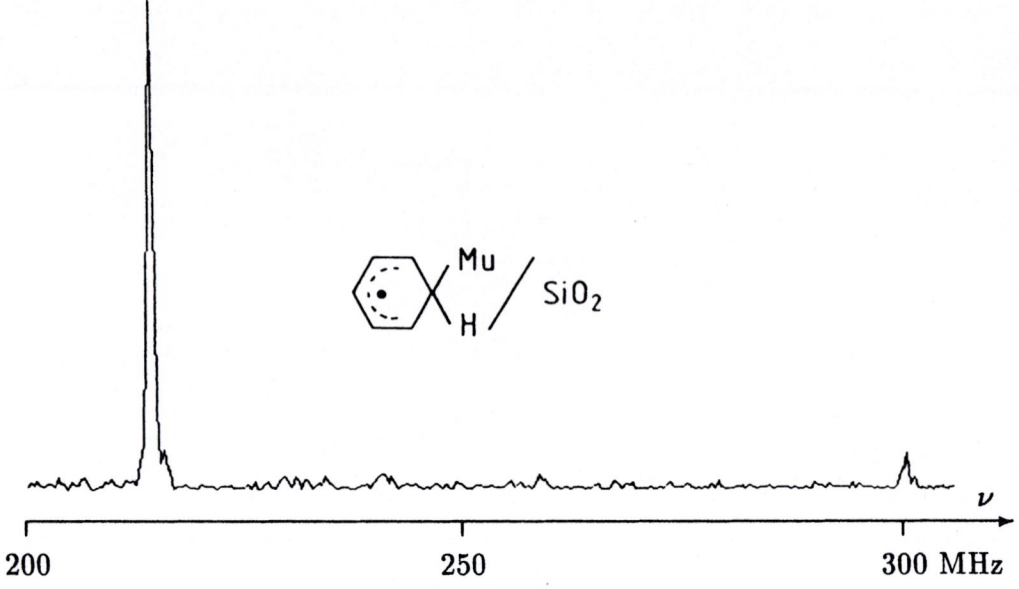

Figure 4.7: Spectrum of the cyclohexadienyl radical adsorbed on a silica surface.

the dynamic behaviour of a muonated allyl type radical on a silica surface [54], but the first observation was again the cyclohexadienyl radical [55]. Figure 4.7 shows the room temperature spectrum. It was obtained using a fine SiO_2 powder with an adsorbed monolayer of benzene. The radical lines are significantly broadened compared with liquid phase spectra owing to the restricted mobility. The higher frequency carries much less muon polarization than its low frequency counterpart. This is the typical situation of loss of coherence due to the finite lifetime of the Mu precursor (see Section 3.2.2). A small increase of the hyperfine coupling constant compared with its liquid phase value is ascribed to the radical/surface interaction, as in the case of the allyl radical.

Table 4.3: MNDO results for $C-C$ bond lengths in nm

	r_{12}	r_{23}	r_{16}	r_{67}
cyclohexadienyl	0.1370	0.1426	0.1504	
benzyl	0.1397	0.1408	0.1437	0.1399

4.3 Structure and hyperfine couplings

4.3.1 Structure of cyclohexadienyl and benzyl

An understanding of hyperfine coupling constants, and of isotope and substituent effects on it, first requires a discussion of the structure. It is useful to do this by comparison with the benzyl radical.

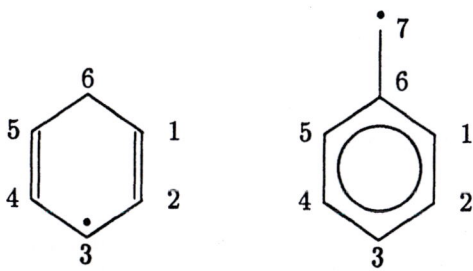

The structures were calculated using the MOPAC program [56] based on the semiempirical MNDO Hamiltonian [57]. It is known to give fairly reliable results for geometries, whereas INDO is still superior for the calculation of hyperfine coupling constants. As expected, MNDO predicts C_{2v} symmetry for both radicals. This leads to wave functions of σ and π symmetry, which is convenient for discussions in terms of *inductive* and *mesomeric* effects. The $C-C$ bond lengths are given in Table 4.3. The values oscillate about those of benzene (0.1407 nm), but the alternation effect is obviously much weaker for benzyl than for cyclohexadienyl. C_1-C_6 is the longest bond in both cases, but in cyclohexadienyl it is still considerably shorter than a $C-C$ single bond (0.1539 nm in $c-C_6H_{12}$). It is expected that this contributes to the unusually high coupling constant of the methylene protons.

4.3.2 Wave functions and hyperfine coupling constants

We are used to looking at the cyclohexadienyl radical as a 5-π electron system. However, it has been recognized that the two methylene hydrogens conjugate with the ring since a negative combination of the corresponding $1s$-AOs forms a p_z-like orbital [58]. Together with the p_z of C_6 this leads

39

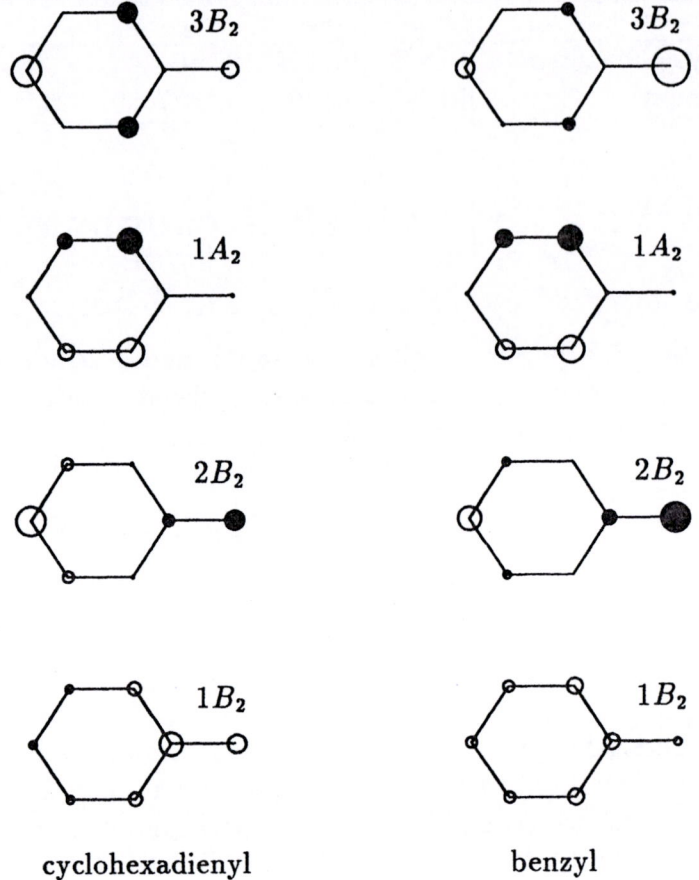

Figure 4.8: INDO wave functions for occupied π-MOs with spin α.

to a 7-π system and makes it isoelectronic with benzyl. If this is a realistic picture we should expect an analogous behaviour of the two radicals, e.g. upon substitution (see Section 5.2).

Figure 4.8 displays the wave functions of the occupied π-MOs with spin α. They were calculated with the INDO program based on the MNDO geometry in Table 4.3. The size of the circles represents the square of the coefficients of corresponding AOs. For the methylene hydrogens of cyclohexadienyl they represent the sum of the two atoms.

For each α-MO there is a corresponding β-MO which is slightly higher in energy but looks quite similar (not shown). The β-MO of symmetry $3B_2$ is empty since there is an odd number of electrons. This makes α-$3B_2$ the leading orbital for spin density distribution on the radical. We note that it has a nodal plane close to the nuclei in *meta* position. Of the ring carbons, the coefficient is higher in *para* than in the *ortho* positions. The highest coefficient for benzyl is on C_7, as expected. A smaller, but still considerable fraction of spin is predicted to populate the methylene protons of cyclohexadienyl.

The remaining orbitals are polarized by the leading orbital. The UHF calculation accounts for this by generating the difference between the α and β MOs in such a way that it enhances the existing α spin population on C_1, C_3, C_5, and $C_7(H_2)$ at cost of the neighbouring even-numbered atoms. This leads to negative spin populations on C_2, C_4, and C_6. This is true for the σ as well as for the π MOs. The methylene protons of cyclohexadienyl demonstrate the relative importance of the contributions: 71.5% of their spin population (and thus of their hyperfine coupling) stems from $3B_2$, 10.4% from the remaining π-orbitals, and 18.1% from the σ system. This cross talk between π and σ system blurs the distinction between inductive and mesomeric substituent interactions, and it has the complicating consequence that substitution on the ring influences the coupling constants via substituent interaction on *all* orbitals, not just on $3B_2$ (*vide infra*).

The hyperfine coupling constants calculated for different geometries by INDO are compared with the experimental values in Table 4.4. For the *meta*-protons the predicted couplings are too high. This is a known deficiency of INDO for protons attached to carbon atoms with negative spin population. High values are also calculated for the methylene protons of cyclohexadienyl, but the values improve considerably with the level of approximation for the geometry. A similar improvement is seen for the *ortho*- and *para*-protons. This is good evidence that the original standard structure [59] is not a realistic approximation, and that the MNDO geometry with its alternating bond lengths represents an improvement. For benzyl, the deviations are smaller, but the alternation effect was also less pronounced.

4.3.3 Isotope effects

Comparison of the hyperfine coupling constants of isotopic cyclohexadienyl radicals (Table 4.1) reveals considerable isotope effects for the nuclei in the methylene group. Quantum chemical calculations for rigid radicals cannot account for this since the electronic wave function is independent of isotopic substitution within the BO approximation. A contribution due to non-BO behaviour is not excluded, but the important part is considered to arise from the mass dependence of the internal dynamics [33]. Two approaches for a quantitative interpretation have been taken so far. The first one calculates

Table 4.4: INDO proton hyperfine coupling constants

	ortho	*meta*	*para*	C_6/C_7	ref.
cyclohexadienyl					
experiment	(-)25.3	7.8	(-)37.1	134.1	[46]
standard	- 31.1	14.3	- 27.5	273.5	[59]
INDO-optimized	- 30.5	15.2	- 31.1	282.7	[46]
MNDO-optimized	- 28.6	12.5	- 33.3	193.1	
benzyl					
experiment	(-)14.4	4.8	(-)17.2	45.7	
standard	- 17.9	10.1	- 15.7	47.6	[59]
INDO-optimized	- 20.1	11.7	- 18.7	48.6	[60]
MNDO-optimized	- 20.0	12.0	- 18.8	45.8	

the hyperfine couplings as an average over all vibrational wave functions in the harmonic approximation, using a MINDO/3 energy hypersurface. It yields factors of 1.16 and 1.00 for the isotope effects of methylene Mu and H in C_6H_6Mu [61]. Considering the fact that they represent sums of positive and negative contributions over all 33 normal modes the agreement is quite satisfactory and shows at least that dynamics can give an effect of the observed magnitude.

The second approach takes anharmonicity into account but does not do the dynamic averaging. The lighter isotope results in a higher zero-point vibrational energy and thus in a longer bond and a larger coupling constant. For a Morse potential [62] with parameters typical for $C - H$ bonds the expectation value for the bond length in the vibrational ground state is 4.9% higher for $C - Mu$ than for $C - H$, whereas for $C - D$ it is only 0.64% shorter [63]. This may even underestimate the effect slightly, since it is known from microwave spectra that in deuterated methyl halides the $C - D$ bond is shorter than $C - H$ by 0.81% (0.0009 nm) [64]. The bond length effect may be simulated approximately in a forced MNDO calculation where one methylene $C - H$ bond is lengthened by 4.9%. The electron distribution adapts to the new nuclear positions. This results in a geometry relaxation on the unchanged electronic energy hypersurface. In the calculation it is taken into account by optimization of the remaining geometry parameters. The main consequence is a shorter second methylene $C - H$ bond by 0.3% and also shorter $C_1 - C_6$ bonds by 0.25%. Except for the methylene group the symmetry plane is nearly retained. The INDO proton coupling constants obtained with this geometry are increased by 25% at the longer and decreased by 4.9% at the shorter methylene $C - H$ bond. They compare favourably with the experimental numbers of +20.5% and -5.9% for the corresponding muon and proton couplings in C_6H_6Mu (Table 4.1). No significant effect is predicted for the *ortho* and *meta* protons, but for the *para* proton of the muonated species a coupling constant is calculated which is less negative by 1.2% . This is again in qualitative agreement with observation. It seems that the enhanced bond length to the lighter isotope gives the key to an understanding of the mass-dependent coupling constants.

It should be noted in this context that MNDO also predicts a 10% larger dipole moment for the muonated case with the increased $C-H$ bond length. This could have kinetic implications, in particular in polar solvents.

Chapter 5

Substituent effects on hyperfine coupling constants

5.1   Monosubstituted radicals

27 monosubstituted benzenes ØX were measured either as pure liquids or in concentrated solutions. We use toluene as a model case to discuss the observed features and the assignment procedure. Three spectra are shown in Figure 5.1. At 3 kG three pairs of lines are observed (top). They correspond to radicals with coupling constants of $A'_\mu = 153.8$ MHz, 155.9 MHz, and 160.0 MHz, i.e. within 5% the same as the value of 161.7 MHz obtained for C_6H_6Mu in benzene. This is obviously typical for cyclohexadienyl type radicals. As was demonstrated for C_6H_6Mu one expects line splitting in a field of 1 kG when the second methylene hydrogen is H, and no splitting when it is D. This is seen for o-dideutero toluene (middle) where the absence of splitting of the two intense lines identifies the *ortho* Mu adduct. For *p*-deutero toluene (bottom) the smallest two lines are unsplit and therefore assigned to the *para* adduct. By default, the lines of intermediate intensity are assigned to the *meta* isomer. The *ipso* adduct is not observed. Because it would have no H in the methylene position it would not show splitting in a field of 1 kG.

We see that the substituent effect on the hyperfine coupling constant is the lowest for the *meta* isomer. This is reasonable since we have seen that the leading orbital for spin density distribution ($3B_2$ in Fig. 4.8) has a node near the *meta* position. Furthermore, we note that the ESR spectrum of *ortho*-methyl cyclohexadienyl has been observed [65], and the substituent shift is found to be the same in ESR as in μSR.

The distribution of polarization between the *ortho*, *meta* and *para* positions corresponds to 48%, 35% and 17%, respectively, which is nearly the same as for practically thermal tritium [66]. The distribution is near statistical. This indicates that there is no big isotope effect in the addition of hydrogen isotopes to toluene, and probably to other arenes as well.

Based on the results with toluene we suggest the following generalized principles for the assignment of the radicals in the other monosubstituted benzenes:

- The *meta* isomer shows the smallest substituent effect.

- The *para* isomer shows the weakest signal.

Wherever possible, the assignment obtained this way is checked by deuteration, by comparison of substituent shifts with those obtained from ESR measurements or of selectivity data with those for H atom addition.

Further typical examples of spectra obtained with monosubstituted benzenes are shown in Figure 5.2. The broken lines give the position of the radical signals in unsubstituted benzene. They visualize the line shifts induced by substitution. We see that for these strongly interacting substituents the shifts can be quite large. The figure demonstrates the power of the μSR method in allowing the simultaneous observation of a number of isomeric radicals. The corresponding ESR spectrum of such a mixture of radicals would consist of more than 100 lines. It would therefore be difficult to analyze.

Substituent effects on the spin population ρ_α of the carbon p_z-orbital of $\cdot CH_3$ are often described by [67]

$$\rho_\alpha = \Pi_X(1 - \Delta_X) \tag{5.1}$$

where the product goes over all substituents. By analogy, we write for the methylene muon coupling of substituted cyclohexadienyl radicals

$$A'_\mu = A'^0_\mu \Pi_X(1 - \Delta_X) \tag{5.2}$$

with $A'^0_\mu = 161.7\,\text{MHz}$, the value for the unsubstituted radical. The difference in comparison with the previous equation is the dependence of Δ_X on the position of substitution. An equivalent parameter has been defined for the benzylic α-hydrogen couplings in benzyl radicals [68], where it was denoted σ_α in analogy to Hammett free energy parameters.

The complete set of muon hyperfine interaction constants on the 73 distinguishable cyclohexadienyl radicals is collated in Table 5.1. They are displayed in the order of increasing values for the *ortho* isomer. For detailed support of the assignments we refer to the original work [69,70].

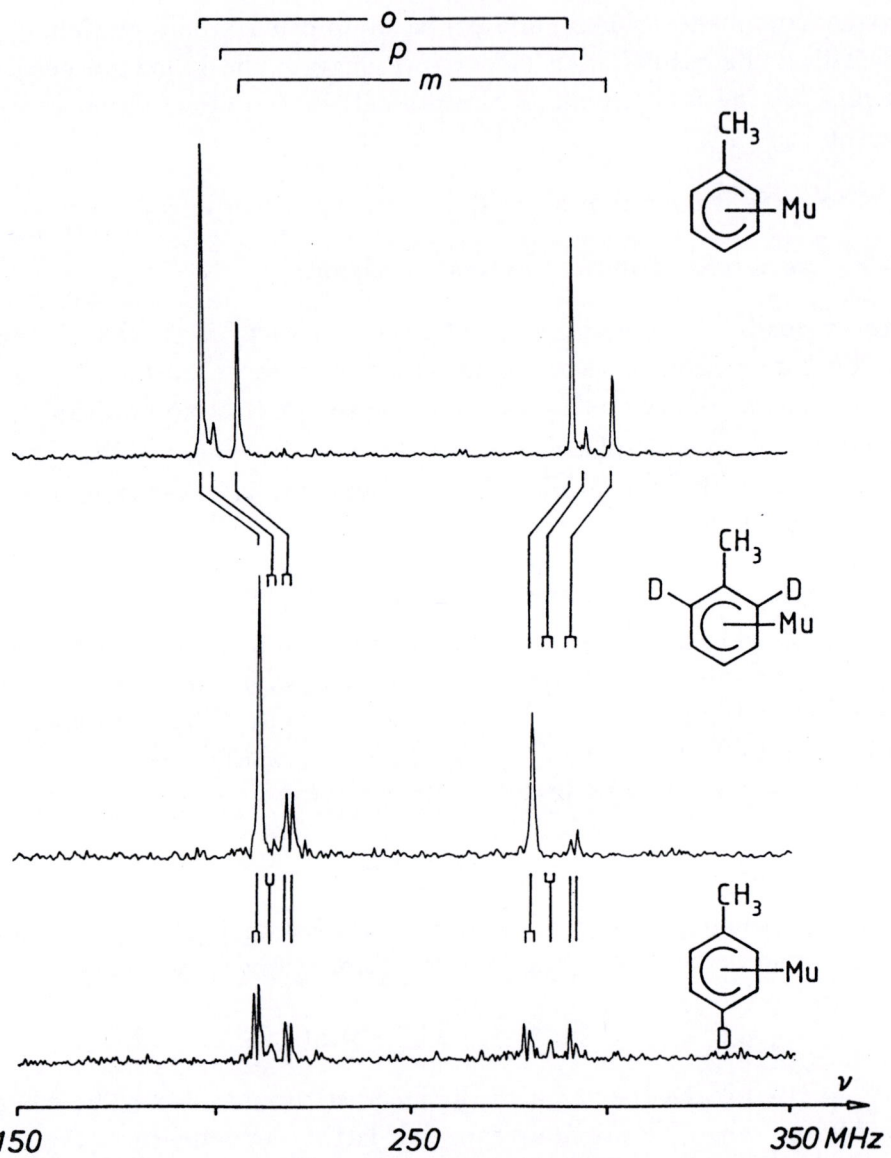

Figure 5.1: Fourier power μSR spectra obtained with toluenes. Top: undeuterated toluene, 3 kG. Middle: o-dideuterotoluene, 1 kG. Bottom: p-deuterotoluene, 1 kG.

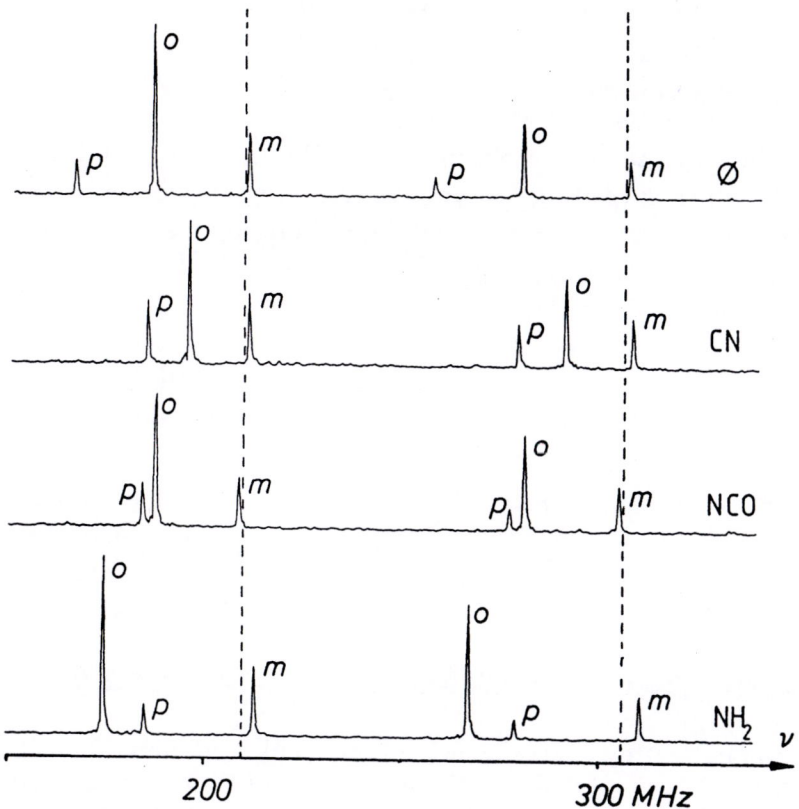

Figure 5.2: μSR spectra obtained with biphenyl, benzonitrile, phenyl iso-cyanate, and aniline at 3 kG.

47

Only in one case, for X = CF$_3$, there was evidence for the formation of a small amount of an *ipso* adduct. For three substituents, X = Br, CHCl$_2$, and CCl$_3$, there are indications that the *ortho* and *para* isomers appear degenerate in the spectrum. For X = F, degeneracy of the *meta* and *para* isomers is resolved by deuteration.

Further inspection of the table reveals that *meta* substitution usually causes a small decrease of the muon coupling. Only six substituents give an increase with a corresponding negative Δ_X value. This is the case mostly for strongly conjugating and therefore polarizable substituents. The majority of Δ_X values is larger for the *ortho* than for the *para* position. The others are marked with an asterisk in the table. They correlate with very small or negative values in the *meta* column, except for X = Cl, CH$_2$Cl. Very large effects with values exceeding 8% are found exclusively for substituents with their own delocalized π system or with non-bonding electron pairs available for interaction with the ring. This is reasonable since it allows delocalization of the α spin onto the substituent. The fact that the effect is greater in certain cases in the *ortho* and in others in *para* position is less plausible. The explanation must be sought in the interaction of *all* substituent orbitals with *all* ring orbitals. In terms of perturbation theory the interaction is particularly strong when the coefficient of the AO at the substitution site is large, and when the MOs are close in energy. The latter is demonstrated nicely with X = CF$_3$. Its orbitals of π symmetry are very low in energy and don't match the ones on the ring. This leads to small substituent effects.

5.2   Comparison with data for benzyl

A suggestion in Section (4.3) that cyclohexadienyl and benzyl radicals should be regarded as isoelectronic species was supported by the similarity of their π-MOs in Figure 4.8. Here we seek experimental evidence for it.

Figure 5.3 shows a plot of the Δ_X values for the benzylic α-hydrogen couplings of benzyl obtained by ESR with those of the cyclohexadienyl methylene muons from Table 5.1. The ESR values are averages from different sources [61,68,71-75]. Some of the values scatter considerably, and obviously erroneous determinations were not used. With the exception of the point for o-F we find a clear correlation. This is particularly noteworthy since the benzylic protons measure the in-plane spin density which occurs via spin polarization of the σ-skeleton whereas the cyclohexadienyl muon couplings are primarily due to the direct π spin density. It reflects the validity of McConnell's relationship on the proportionality between spin density at in-plane protons

Table 5.1: Substituent effects on hyperfine coupling constants of mono-substituted cyclohexadienyl radicals, in units of $100\Delta_X$. The asterisk marks the cases where Δ_X is larger for the *para* than for the *ortho* position.

substituent	ortho	meta	para
CF_3	2.78	0.79	1.15
$COOCH_3$	2.99	−0.08	*8.78
$COOH$	3.40	0.12	*9.09
$C(CH_3)_3$	3.69	1.20	2.37
$CH_2CH=CH_2$	4.20	1.09	
CH_2OH	4.39	1.13	3.61
$OCOCH_3$	4.39	0.27	2.22
$CH_2\emptyset$	4.70	1.22	3.21
I		1.28	
Br	4.72	1.17	
CH_3	4.86	1.03	3.54
CH_2Cl	5.31	1.54	*6.26
Cl	5.31	1.01	*5.69
F	5.62	0.54	0.54
CN	5.62	−0.70	*9.99
$CHCl_2$	6.28	2.66	
CCl_3	7.21	3.69	
$O\emptyset$	8.16	0.58	4.59
$C\equiv CH$	8.51	−0.51	*17.04
SH	9.06	0.74	
NCO	9.15	0.33	*10.51
OH	9.17	−0.54	4.20
$\emptyset$	9.85	−0.16	17.98
OCH_3	11.82	1.77	4.43
$COCl$	12.51	0.87	8.78
NH_2	14.24	−1.69	9.95
$CH_2=CH_2$	17.43		*26.53

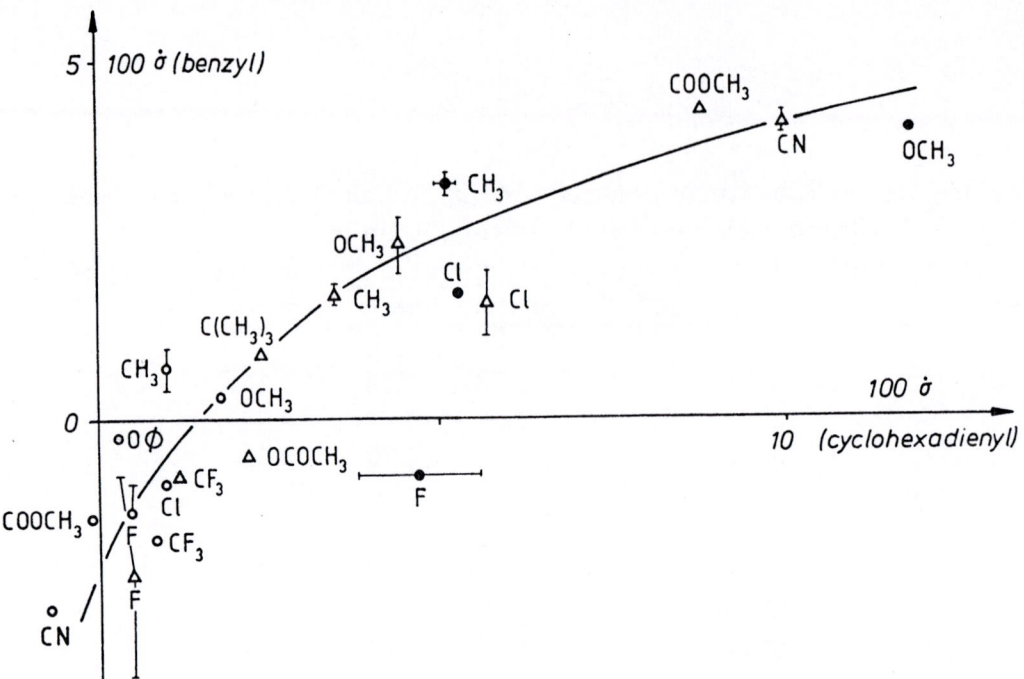

Figure 5.3: Correlation of substituent Δ_X-values of benzyl vs. cyclohexadienyl radicals for *ortho* (•), *meta* (o) and *para* ($\triangle$) substitution.

and π spin population at adjacent carbon atoms [76], and it supports the isoelectronic character of the two species.

5.3 Polysubstituted radicals

5.3.1 Experimental observations

A series of polysubstituted cyclohexadienyl radicals was also investigated [33,77]. A typical spectrum is shown in Figure 5.4. It was obtained with *m*-xylene. Apart from the signal corresponding to muons in diamagnetic environments (D) and a small background signal (Cy) on the cyclotron frequency we find four pairs of lines corresponding to radicals ($R_1 - R_4$) with reduced coupling constants of $A'_\mu = 146.5$ MHz, 149.0 MHz, 152.2 MHz, and 158.3 MHz. From eqn. (5.2), using the Δ_X values for CH_3 in Table 5.1 one obtaines 146.3 MHz for 1,5 substitution, 148.4 MHz for 1,3 substitution, and 158.3 MHz for 2,4 substitution. This is in good agreement with three of the four observed values and allows immediate assignment. It is gratifying to note that the strongest lines correspond to 1,3 substitution which is degenerate with the 3,5 isomer and therefore statistically favoured over the other

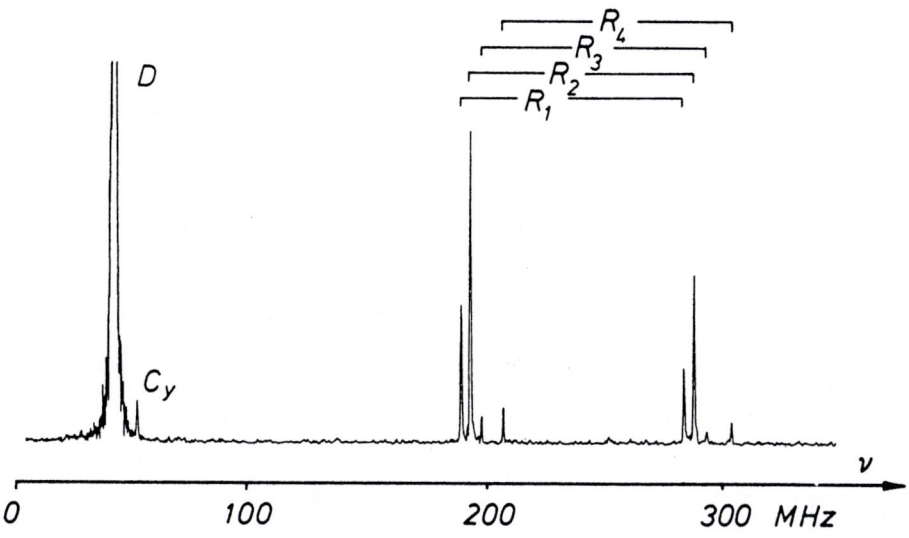

Figure 5.4: Fourier power μSR spectrum obtained with m-xylene in a transverse field of 3 kGauss.

isomers. The weakest lines (R_3 in Figure 5.4) must belong to the *ipso* Mu adduct which was not observed with toluene, but which is also enhanced by a statistical factor of two in m-xylene.

The experimental coupling constants are plotted *versus* the ones calculated based on eqn. (5.2) in Figure 5.5 (o) for all six doubly and six higher methyl substituted radicals. We note that most of the calculated values are low. We therefore propose to use the relation

$$A'_\mu = A'^0_\mu \Pi_X (1 - \Delta_X) \Pi_{XY} (1 - \Delta_{XY}). \tag{5.3}$$

This differs form eqn. (5.2) by an additional correction term Δ_{XY} which reflects a position dependent interaction between two substituents X and Y. Δ_{XY} values are calibrated on disubstituted radicals. This brings the points onto the 45° line by definition (• in Figure 5.5). They are used to calculate new values for three- and four-fold substituted cases (⊙). It is obvious that this improves the situation and brings all calculated values to ≤ 1 MHz within the experimental numbers.

A similar plot can be obtained for fluorinated cyclohexadienyl radicals. Even crowded species such as tetramethyl and pentafluoro radicals are described well with eqn. (5.3). This indicates that they are probably still planar.

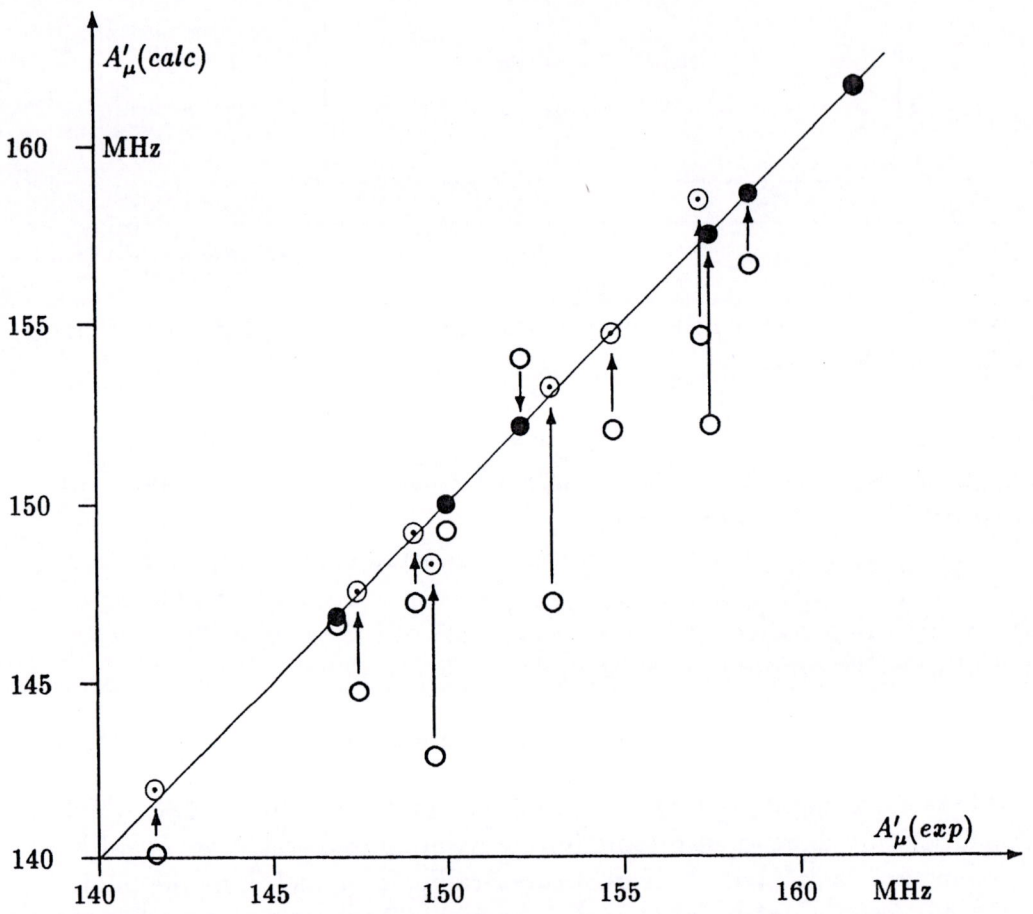

Figure 5.5: Calculated *vs.* experimental muon hyperfine coupling constants in poly-methyl substituted cyclohexadienyl radicals (see text for the meaning of the symbols).

Table 5.2: Parameters for the interaction of two substituents on cyclohexadienyl, in units of $100\Delta_{XY}$

position of X,Y	$X = CH_3$ $Y = CH_3$	F F	Cl Cl	OCH_3 OCH_3	OCH_3 CN	CN OCH_3
$1,5 \equiv o,o'$	-0.13	-1.93	-0.24	-2.62	$+2.62$	$+2.62$
$1,2 \equiv o,m$	-1.81	-1.62	-0.46	-7.07	$+0.72$	-2.60
$1,4 \equiv o,m'$	$+0.36$	$+1.12$	-0.07	$+4.44$	-1.20	-2.96
$1,3 \equiv o,p$	-0.45	-0.71	$+0.04$	-3.37	$+5.17$	$+6.98$
$2,4 \equiv m,m'$	$+0.01$	-0.90	-0.40	-1.45	-1.18	-1.18
$2,3 \equiv m,p$	-0.99	$+1.53$	$+0.08$	-2.53	-2.59	-1.37

5.3.2 Do coupling constants reflect *capto-dative* stabilization?

The Δ_{XY} values for doubly methyl substituted radicals are collected in Table 5.2 together with those for doubly F, Cl, OCH$_3$ and mixed OCH$_3$/CN substitution.

Viehe and coworkers [78] postulated that the combined action of acceptor and donor substituents on a radical center should lead to an enhanced stabilization compared to acceptor/acceptor or donor/donor substitution. Meanwhile, the concept has been applied successfully to organic synthesis involving radicals [79].

It is well accepted that electron delocalization has a stabilizing effect. It has therefore been postulated that spin delocalization also reflects stabilization. Indeed, correlations of hyperfine coupling constants with Hammett σ parameters [80] or with rate constants of reactions involving radicals or transition states with radical character [68] were found. Methyl proton coupling constants in 3,7- and 7,7-cyano/methoxy substituted benzyl radicals showed clearly the behaviour which is expected based on the concept of *capto-dative* substitution [81,82].

In terms of eqn. (5.3) one expects negative values for Δ_{XY} when X and Y are both donors or both acceptors, in particular when they are identical, and positive when one is a donor and the other one is an acceptor. Inspection of Table 5.2 reveals that this holds true for 1,3- and for 1,5-substitution (except for the small positive value for Cl). It is almost consistently wrong for the 1,4-substituted cases. For the remaining cases where one or both of the substituents occupies a *meta* position the signs scatter and the coupling constants do not reflect the postulated synergistic or antagonistic expectations. This does not necessarily mean that the radicals are not capto-datively stabilized. Energy is a matter of electron density distribution. Owing to spin polarization effects spin density distributions can be quite different, in particular in these delocalized systems (replacement of a fraction of α—spin by the same amount of β—spin changes spin but not charge distribution).

It is interesting to note that for the 1,2 and for the 1,4 isomers the effects are quite different, although in either case the substituents occupy the *ortho* and *meta* positions. This is probably not a question of steric interaction and concomitant radical deformation, since for the 2,3 isomers the effect of neighbouring substituents is much smaller. Based on the unperturbed orbitals (Figure 4.8) this left-right assymmetry can be obtained only by admixture of $1A_2$ character to the leading orbital for spin distribution, $3B_2$.

The smallness of the Δ_{XY}-values for Cl are remarkable. Apparently, two chlorine atoms have almost no interaction across the ring.

Chapter 6

The process of radical formation

6.1 Routes to the cyclohexadienyl radical

Possible distribution processes of muons between different chemical states at the end of its thermalization track were sketched in Section (1.3.3) and in Figure 1.1. We now investigate them more in detail for benzene. Formally, the cyclohexadienyl radical is derived by Mu addition to the benzene molecule

$$Mu + C_6H_6 \longrightarrow \cdot C_6H_6Mu. \qquad (6.1)$$

Here, the direct muonated radical precursor is Mu, either hot or thermal. Alternative routes involve diamagnetic muonated precursors, as in

$$Mu^+ + C_6H_6 \longrightarrow [C_6H_6Mu]^+ \xrightarrow{e^-} \cdot C_6H_6Mu, \qquad (6.2)$$

or

$$e^- + C_6H_6 \longrightarrow [C_6H_6]^- \xrightarrow{Mu^+} \cdot C_6H_6Mu. \qquad (6.3)$$

Mu itself could also be formed in an end-of-track process,

$$Mu^+ + e^- \longrightarrow Mu \xrightarrow{C_6H_6} \cdot C_6H_6Mu. \qquad (6.4)$$

Different types of experiments have been devised to test these models. Selectivity measurements in competition reactions should distinguish between hot and thermal Mu, scavenging experiments should probe the influence of radiolytical end-of-track effects, and investigation of the degree of transfer of muon polarization via amplitude and phase effects should discriminate between paramagnetic and diamagnetic precursors and set bounds on the time scale of the formation process.

6.2 Radical formation in cyclohexane solution

In pure cyclohexane 20% of the muons form Mu [83]. For binary mixtures of benzene in cyclohexane one therefore expects to observe cyclohexadienyl radicals with a Mu precursor. Amplitudes and phases of the radical lines should show the behaviour calculated in Section (3.2.2).

It is difficult to determine absolute phases since the experimental time zero has to be known with an accuracy of a fraction of a nanosecond. We therefore use only the phases Φ_r relative to those in neat benzene. They are shown in Figure 6.1 together with the polarizations. There is a clear increase in the phase shift as the concentration of benzene decreases, which is due to the increased precursor lifetime. The direction of the shifts reveals that the muonated precursor was paramagnetic with a hyperfine coupling constant exceeding the one of the radical, i.e. in accord with expectation for Mu. In pure benzene, the polarizations of the two lines are the same within the experimental error, while for lower concentrations the higher frequency carries less polarization.

Quantitative analysis was done by a fit to the theoretical expressions simultaneously to the Φ_r values and to the ratio of the two $|P^{nm}|$ values at each composition. The only free parameter was the bimolecular rate constant for Mu addition. We obtained $k_{Mu} = 8.9(6) \cdot 10^9 M^{-1}s^{-1}$. This is in the range of the values found in direct determinations for different solvents [84] and thus very reasonable. Furthermore it is clearly below the diffusion controlled limit which is expected to be about the same as in water, i.e. $3 \cdot 10^{10} M^{-1}s^{-1}$ [2]. This fact does not promote the idea of a hot Mu precursor. Extrapolated to pure benzene we obtain 10 ps for the lifetime of Mu, which leads to a loss of polarization due to slow formation of only 2.5%. h_R was calculated for each set of P_R values and plotted in Figure 6.1 (broken line). Extrapolation to neat cyclohexane gives a value of 0.18, which is in good agreement with the muon polarization observed directly as Mu, P_M=0.20.

Myasishcheva et al. [85,86] analyzed P_D values for aromatic compounds and their binary mixtures with different saturated molecules in terms of a kinetic model. They assumed Mu as the only and common precursor of muonated species. The rate constants which they deduced for Mu addition are not unreasonable (e.g. $3.1 \cdot 10^9 M^{-1}s^{-1}$ for benzene), but for abstraction reactions they come out orders of magnitude too high (e.g. $2.4 \cdot 10^9 M^{-1}s^{-1}$ for H abstraction from $c\text{-}C_6H_{12}$, where the direct observation of Mu in this compound has now revealed $k \leq 1.6 \cdot 10^5 M^{-1}s^{-1}$ [83]). Their model of thermal Mu as the general precursor also for diamagnetic compounds is thus clearly

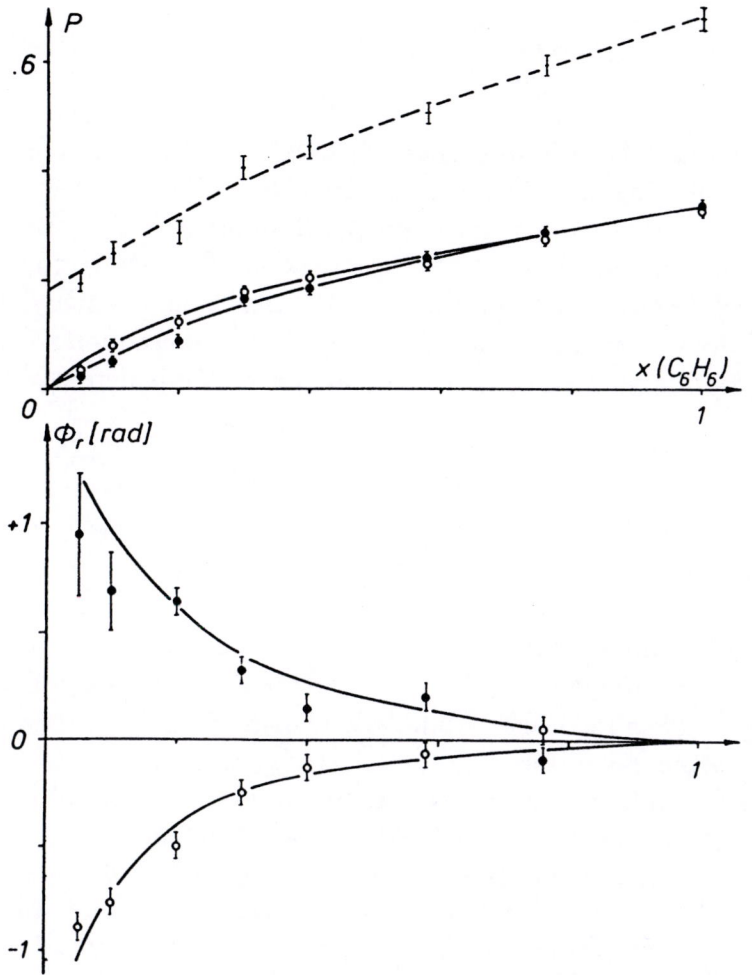

Figure 6.1: Muon polarizations and relative phases corresponding to the lower frequency (○) and the higher frequency (●) of the muonated cyclohexadienyl radical in benzene/cyclohexane mixtures.

inadequate.

6.3 Selectivity in mixtures of benzene and dimethylbutadiene

Figure 6.2 displays the distribution of muon polarization in binary mixtures of benzene with 2,3-dimethylbutadiene (DMBD). Both compounds form a radical, and we assume that they compete for Mu as a common precursor. P_D shows only a weak dependence on composition, whereas the strong curvatures in P_R reveal a high selectivity in favour of radical formation from DMBD. In mixtures with a mole fraction $x_{DMBD} > 0.2$ the cyclohexadienyl radical is not observed because of its broad lines due to chemical reaction with DMBD (see Section 8.5). The selectivity of the reaction of Mu is defined by

$$S_{Mu} = \frac{k(Mu + DMBD)}{k(Mu + C_6H_6)} = \frac{P_R(DMBDMu) \cdot x_{C_6H_6}}{P_R(C_6H_6Mu) \cdot x_{DMBD}}. \tag{6.5}$$

There are two positions where Mu adds to DMBD but six in the case of benzene. Statistically S_{Mu} should be 1/3. Experimentally it is found to be ≈ 5, with a slight dependence on composition [88]. If we use the rate constant derived above for Mu addition to benzene we obtain $4 \cdot 10^{10} M^{-1} s^{-1}$ for Mu addition to DMBD, which is certainly close to the diffusion controlled limit in these solvents. The same rate constant was measured for the reaction at room temperature in the gas phase, but the selectivity with respect to benzene is $S_{Mu} = 65$ [87]. This is in line with the results for the corresponding reactions of H where one has $S_H \approx 100$ in the gas phase but only ≈ 10 in aqueous solution [88]. In agreement with the findings for other addition reactions [5,18,89] Mu is faster than H and therefore less selective. Tunnelling favours the lighter isotope in these exothermic reactions. The dramatic solvation effect which appears for Mu and for H is not yet understood.

The analogous behaviour of Mu and H reactions supports the view that Mu is the radical precursor in benzene, and the observed selectivity makes it unlikely that this is highly epithermal Mu.

6.4 Scavenging experiments

6.4.1 The effect of carbon tetrachloride

CCl_4 is an excellent scavenger of electrons. It acts via dissociative capture of the negative charge and may therefore interfere in processes of radiolytical

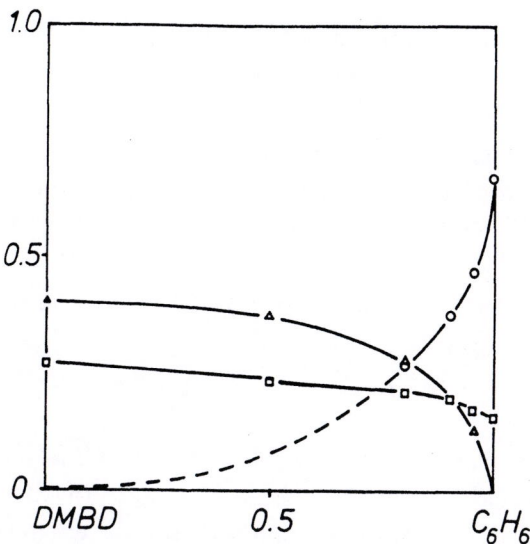

Figure 6.2: Muon polarizations observed with benzene/DMBD mixtures as a function of mole fractions. Muons in diamagnetic environments ($\square$), in the trimethylallyl radical ($\triangle$) and in the cyclohexadienyl radical ($\circ$).

nature (eqns. 6.2-6.4). The system was investigated previously [90], but only P_D was reported. A considerable increase in P_D was observed in ≈ 1 M CCl_4 solution. Based on qualitative arguments the adequacy of a radiolytical model was rejected. Inhomogeneous scavenging is not a linear function of scavenger concentration but relatively more important at low concentrations. We therefore investigated in particular the region ≤ 1 M and measured both, P_D and P_R. The results are shown in Figure 6.3. We find a significant effect even at the lowest concentration, 0.01 M, and a smooth decrease of P_R and $1 - P_D$ with increasing concentration. This shows that CCl_4 interferes strongly in the distribution process of muons between radicals and diamagnetic species. It reacts efficiently with electrons but only slowly with Mu ($k \approx 5 \cdot 10^8 M^{-1}s^{-1}$), furthermore it does not react with cyclohexadienyl radicals ($k < 10^5 M^{-1}s^{-1}$). If we accept that Mu is the direct radical precursor we must conclude that CCl_4 inhibits Mu formation by scavenging spur electrons before their combination with Mu^+, i.e. model (6.4) is an adequate description of radical formation in benzene.

For quantitative analysis we apply the empirical expression

$$h_R = \frac{h_R^0}{1 + (\alpha[Sc])^\beta},\qquad(6.6)$$

which has been used to describe inhibition of o-positronium [92]. It is related to the *Warman-Asmus-Schuler* scavenging function [93]. h_R^0 and h_R are the fractions of muons ending up in radicals in neat benzene and in solutions with scavenger concentration [Sc], respectively. α is the scavenger reactivity and β is an adjustable parameter which is related to the distribution function of reactive species in the spur and was originally assumed to be 0.5 [93]. Equation (6.6) does not extrapolate correctly to neat scavenger. It is usually applied for $[Sc] < 0.5$. We obtain excellent fits for $CCl_4 \leq 1$ M, first assuming $h_R = P_R$, and secondly assuming $h_R = 1 - P_D$ since it had been suggested by analogy to results in the aqueous phase [91] that the missing fraction P_L corresponds most likely to muonated radicals which lost spin polarization during encounters with other paramagnetic species near the end of the muon track [89].

For $h_R = P_R$ we obtain $\alpha = 1.2(1)$ M^{-1} and $\beta = 0.53(3)$, for $h_R = 1 - P_D$ we have $\alpha = 0.42(4)$ M^{-1} and $\beta = 0.82(6)$. None of the interpretations is favoured at this point. In the first case β agrees with the original value, 0.5, in the second case with the literature value for o-Ps inhibition in the CCl_4/benzene system, 0.93(5) [92]. The inhibitor constant α is at least an order of magnitude lower than that found for electron scavenging in cyclohexane ($\alpha = 12$ M^{-1}) or for Ps inhibition in benzene ($\alpha = 17.4$ M^{-1} [92]). Both, μ^+ and e^+, probe the end of their ionization tracks. The difference in α may mean that the last spurs are different in the two cases or that the two particles come to rest at a somewhat different position with respect to the centre of the last spur.

The lifetime of Mu is about the same in these mixtures as in neat benzene. We thus have almost complete transfer of muon polarization, as shown above, and in the absence of competing channels for Mu we have $h_R \approx h_M$. We can therefore estimate the time scale of Mu formation using

$$\alpha = k_e/\lambda, \tag{6.7}$$

where k_e is the bimolecular rate constant for the reaction of CCl_4 with electrons and λ characterizes the recombination rate of the ions in the pure hydrocarbon [94]. Assuming that k_e is of the same order as in cyclohexane ($k_e = 2.7 \cdot 10^{12}$ M^{-1}s^{-1}) we determine $\tau = 0.6/\lambda$ where half of the ions have recombined. This reveals that Mu must be formed within about one picosecond after creation of the last spur. This is the same result as that obtained previously for the aqueous system [95]. It is of course important that $\tau \ll \omega_0^{-1} = 36$ ps since otherwise an ensemble of Mu atoms would be formed with partly dephased muons which would give rise to a loss of polarization.

The difference between the two curves in Figure 6.3 constitutes the lost fraction P_L of muon polarization. Due to the logarithmic scale it is not so obvious that it exhibits a pronounced maximum near $[CCl_4]=0.3$ M. If we

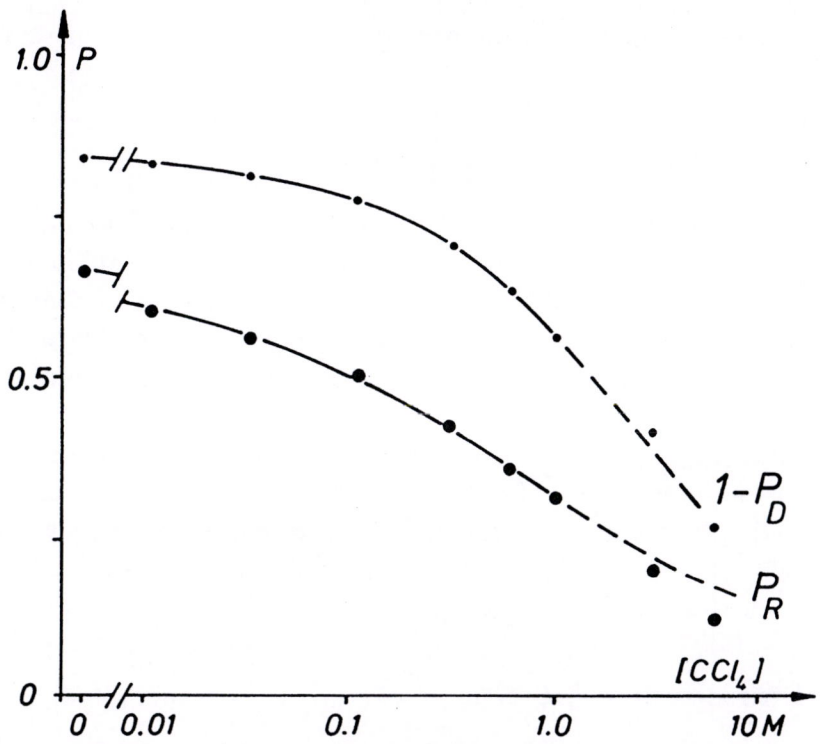

Figure 6.3: Muon polarizations in CCl$_4$/benzene mixtures and fit of eqn. (6.6) for concentrations ≤ 1 M. Note the logarithmic scale.

accept that P_L represents radicals which lost polarization in encounters with other paramagnetic species near the end of the track the following processes can qualitatively explain the maximum: in pure benzene, Coulomb forces furnish the recombination of benzene cations and ionization electrons, thereby reducing rapidly the concentration of paramagnetic species. In low concentrations of CCl_4 the electrons are transformed into diamagnetic Cl^- ions via dissociative electron capture. The remaining radical, $\cdot CCl_3$, is neutral and has therefore a longer lifetime and an augmented probability to become involved in depolarizing encounters with C_6H_6Mu. At higher scavenger concentrations radical formation is increasingly inhibited, which automatically reduces P_L.

6.4.2   The effect of methyliodide

ICH_3 has an effect quite similar to CCl_4 on the muon polarization in benzene [96]. Even at $[ICH_3] = 0.3$ M the μSR lines of the cyclohexadienyl radical are observed unbroadened, which demonstrates the absence of chemical reactions. ICH_3 may react with Mu, but a corresponding rate constant is not available. However, even if this was diffusion controlled, ICH_3 would compete with benzene only at the highest concentrations.

The data were fitted to eqn. (6.6) in the same way as it was done for CCl_4. For $h_R = P_R$ we obtain $\alpha = 1.0(2)$ M^{-1} and $\beta = 0.63(8)$, for $h_R = 1 - P_D$ we have $\alpha = 0.44(5)$ M^{-1} and $\beta = 0.75(8)$. Within error, these values agree with those for CCl_4. The conclusions for the end-of-track processes are therefore consistent with those derived for CCl_4.

6.4.3   The effect of iodine

Iodine is a relatively non-selective but highly efficient scavenger . It reacts at a diffusion controlled rate with electrons and with most radicals, but it is sufficiently inert towards benzene. It was therefore used to convert muonated species in benzene to diamagnetic products under retention of some of the polarization.

The residual polarizations and phases obtained in a transverse field of 200 G for a series of concentrations $[I_2] \leq 0.5$M are shown in Figure 6.4. P_D increases from 0.16 in pure benzene to 0.79 in the highest concentration. This is accompanied with a pronounced phase shift at intermediate concentrations. A similar set of data with however nearly negligible phase shifts was obtained at 3000 G.

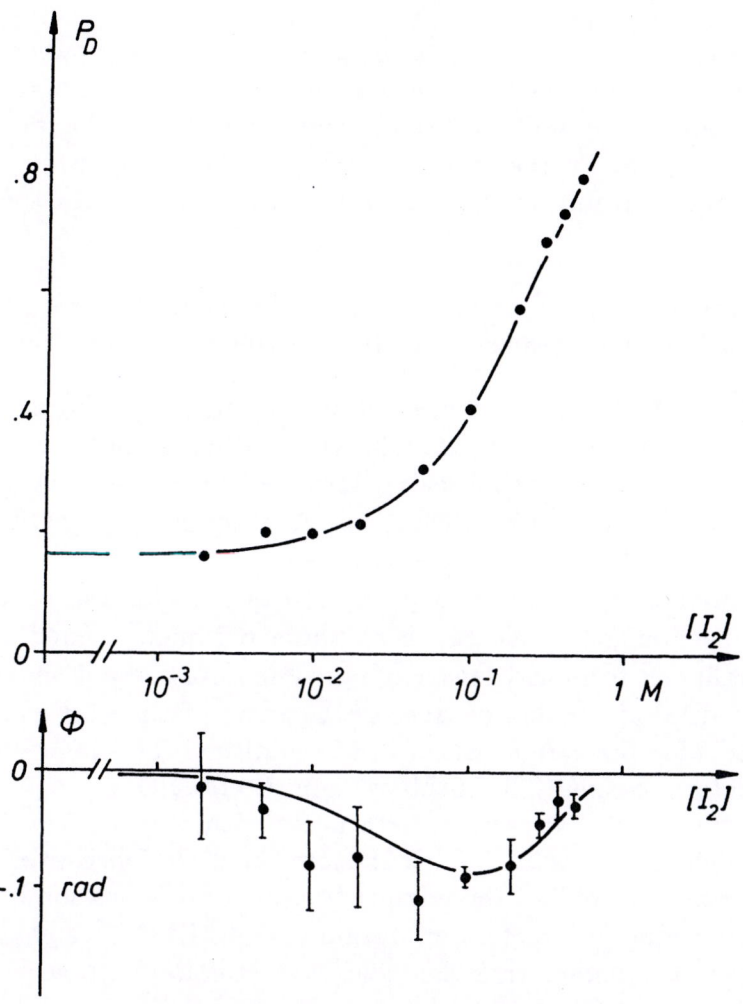

Figure 6.4: Residual diamagnetic muon polarizations and phases for solutions of iodine in benzene at 200 G.

The data were fitted to the model outlined in Section (3.2.3). P_D was assumed to be composed of the following three contributions:

- The prompt fraction, $P_D(0) = 0.16$, which is found in pure benzene and does not change with $[I_2]$. It determines the reference phase, $\phi = 0$.

- A phase shifted fraction h_R with the cyclohexadienyl radical as precursor. It is given by the scavenging relation (6.6). It does not significantly contribute to P_D and ϕ at low concentrations where the radical lifetime τ_R is too long. Furthermore, at high concentrations, τ_R is too short to allow much evolution of polarization and we have nearly full transfer of P.

- The fraction $h_R^0 - h_R$ which because of scavenging does not enter the radical channel. It is concentration dependent but not phase shifted.

The system had been measured previously in the classical residual polarization study by *Brewer et al.* [12] which provided indirect evidence for the formation of muonated radicals. Here we have the advantage of higher counting statistics, but in particular we know the muon hyperfine coupling constant ($A_\mu = 514.6$ MHz, Brewer assumed 424 MHz) and the rate constant for the reaction of I_2 with C_6H_6Mu ($2.5 \cdot 10^9 M^{-1}s^{-1}$ from a direct measurement, see Section 8.3). We can keep these parameters fixed and instead extend the model to include scavenging. The data are not sensitive to the nature of radical precursors because of the short lifetime. *Brewer* obtained $5 \cdot 10^8 M^{-1}s^{-1}$ for the rate constant of Mu addition. This is in disagreement with our determination ($8.9 \cdot 10^9 M^{-1}s^{-1}$, Section 6.2).

The results of the two sets of data agree within error. We obtain $h_R^0 = 0.80(2)$, which is more than $P_R = 0.65$ from the direct observation. It therefore supports the view that the missing fraction of polarization is part of P_R. For the scavenging parameters we obtain $\alpha = 4.14(3)M^{-1}$ and $\beta = 0.99(3)$. α is considerably higher than for CCl_4 but still low compared with typical values obtained for electron tracks. Here, β adopts unity, the value for *Stern-Volmer* kinetics in homogenous systems.

The relative importance of the contributions to P_D from the different channels may be demonstrated for the case of 0.5 M I_2: 0.16 is $P_D(0)$, 0.09 is the contribution transferred from the radical with h_R=0.26, and 0.54 is the scavenged fraction.

Chapter 7

Distribution of muons in substituted benzenes

7.1 Monosubstituted benzenes

7.1.1 End-of-track effects on the muon distribution between radicals and diamagnetic species

Muon polarization values are given in Table 7.1. P_D values range between 0.15 for unsubstituted benzene and 0.76 for thiophenol. They increase in the series X = F < Cl < Br < I and in X = CH_3 < CH_2Cl < $CHCl_2$ < $CHCl_3$, and they are higher for α-chloro toluenes than for chlorobenzene. They correlate with C − Cl bond dissociation energies and parallel the efficiency of the reaction of these compounds with electrons via dissociative capture [97].

Increasing P_D values with increasing chlorination are also found for the vapor phase [98] where they are not associated with end-of-track effects (although there is of course an ionization track). The gas phase results are more supportive of hot atom or even hot ion (μ^+) reactions. However, P_D values are substantially lower in gases than in corresponding liquids, indicating that the processes involved may be quite different. A clarification of these differences may be obtained in experiments with pressurized samples on both sides of the critical point in order to observe a continuous transition between the gas and the liquid phase. We look forward with interest to the results of such experiments which are currently carried out at TRIUMF.

Solids may be a somewhat better approximation to liquids than gases. In the solid phase P_D values are often also lower than in the liquid. For hot processes, such a difference is not expected. End-of-track processes can more easily explain the effect. The sudden change of mobility at the phase transition leads to different conditions for the combination of radiolytical species.

Among the low P_D compounds there are at least two (X = CF$_3$, CN) which react efficiently with electrons [97]. In these cases the electron capture is not dissociative. The anion formed can combine with a muon and produce a radical as in eqn. (6.3).

P_D can also increase via scavenging of positive muons. This is probably the case for X=OH, CH$_2$OH, SH and NH$_3$ which can exchange free muons for protons or solvate them strongly as in aqueous solutions. A correlation of P_D with donor number, a measure of the basicity of a molecule, has been observed previously with mostly non-aromatic molecules [99].

The formation of muonated radicals is in competition with the formation of diamagnetic compounds. Therefore, an increase in P_D should naturally be accompanied with a decrease in $\sum P_R$. Inspection of Table 7.1 confirms this as a general trend. However, $P_D + \sum P_R$ ranges from 0.53 for X=NCO to 0.91 for X=SH, i.e. there is a varying fraction of lost polarization.

7.1.2 Regioselectivity in the formation of radicals

We now focus on the distribution of muon polarization between the *ortho*, *meta*, *para* and *ipso* isomers. *Ipso* addition of Mu was observed only for X = CF$_3$, with very low amplitude. It seems to be generally supressed.

All three isomers have been observed in 19 out of 24 substituted benzenes. In many of these cases *ortho* addition occurs with a somewhat higher than statistical probability, even in the presence of a bulky substituent such as C(CH$_3$)$_3$. Only in the case of X = COCl is *ortho* addition significantly less than 40%. The addition of H and T occurs also preferentially or at least in the statistical ratio at the *ortho* position for many substituents, and no exception was reported (see references given in [69]).

Appreciably less than 20% of the of the total yield were found in the *para* isomer for X = CH$_2$Ø, C(CH$_3$)$_3$, F and OCOCH$_3$, whereas almost 30% were observed for X = COCl. Substituents with the ability to enlarge the π system of the ring seem to depress the formation of the *meta* isomers. It is difficult to find a correlation of these deviations with any substituent properties, as for example with Hammett σ parameters. Rather, they may be related to small energetic differences of the radicals formed, as it was found for methyl substituted butadienes [89]. Furthermore they may have to do with possibly differing radical formation processes as outlined above.

For a discussion of the cases where less than three isomers were observed we refer to the original literature [69].

Table 7.1: Distribution of muon polarization in mono-substituted benzenes

substituent	P_D	$\sum P_R$	distribution of $\sum P_R(\%)$		
			ortho	para	meta
H	0.15	0.65			
NCO	0.18	0.35	45	19	36
F	0.19	0.43	42	14	44
$CH_2\emptyset$	0.20	0.44	48	10	42
$O\emptyset$	0.22	0.51	38	24	38
CF_3	0.24	0.42	37	19	44
CH_3	0.25	0.50	48	17	35
OCH_3	0.25	0.43	45	18	38
$C(CH_3)_3$	0.26	0.41	47	13	40
CN	0.27	0.32	43	25	32
$COOCH_3$	0.30	0.33	46	21	33
Cl	0.33	0.28	39	17	44
NH_2	0.36	0.42	48	22	30
CH_2OH	0.39	0.35	40	24	36
$COCl$	0.41	0.23	31	29	40
$OCOCH_3$	0.43	0.23	44	14	43
Br	0.46	0.30	$58^{a)}$	$a)$	42
I	0.52	0.08	–	–	100
CH_2Cl	0.53	0.27	40	26	34
$CHCl_2$	0.59	0.24	$66^{a)}$	$a)$	34
CCl_3	0.67	0.16	$64^{a)}$	$a)$	36
SH	0.76	0.15	51	–	49
OH	$b)$	$b)$	42	16	42
$COOH$	$b)$	$b)$	46	25	29
$\emptyset$	$b)$	$b)$	46	23	31

$a)$ *Ortho* and *para* isomers possibly degenerate
$b)$ No absolute value, measurement in solution.

There are only a few examples where the regioselectivity of addition of heavier hydrogen isotopes is known quantitatively. *Pryor et al.* [66] measured it for close to thermal T. They found 45.7%, 41.2%, and 13.1% for *ortho*, *meta* and *para* addition to toluene, and about 65%, 20% and 15% in the case of aniline and phenol, whereas the corresponding numbers for benzonitrile are 42%, 45%, and 13%. The selectivity pattern of the two isotopes is thus approximately the same. In general, Mu is slightly less selective. This is in line with expectation [18] since Mu addition is faster and only a factor of five from the diffusion controlled limit. A further reason for deviations may be that T experiments involve chemical degradation for the analysis of the selectivity. Mu measures the product distribution on a much shorter time scale and should be more direct.

7.1.3   Relative rate constants and partial rate factors for Mu addition

Table 7.1 demonstrated the influence of a substituent on the regioselectivity of Mu addition. One would also expect a general activating or deactivating effect. This is observed in binary mixtures of substituted benzenes ØX with the unsubstituted molecule ØH. It is demonstrated here with aniline. Figure 7.1 displays the observed polarizations for different compositions, and it is obvious that there is a slight preference for addition to aniline.

Relative rate constants $k_r(\text{Ø}X)$ are calculated from

$$k_r(\text{Ø}X) = \frac{k(Mu + \text{Ø}X)}{k(Mu + \text{Ø}H)} = \frac{\sum P_R(Mu\text{Ø}X)}{P_R(Mu\text{Ø}H)} \cdot \frac{x}{1-x}, \tag{7.1}$$

where x is the mole fraction of benzene. For aniline we obtain a value of 1.3(1) [100], i.e. NH_2 activates the aromatic compound towards Mu addition. The corresponding numbers for H and Mu addition in aqueous solution are 2.4 [101] and 2.7 [102]. NH_2 is known to be a strong π-electron donor. The fact that it accelerates the rate of H and Mu addition indicates an electrophilic nature of these atoms. The rate constants of H were indeed found to correlate reasonably well with Hammett σ values, giving $\rho = -0.45$ [101] (the small negative value indicates electrophilicity), but for Mu $\rho = +0.6$ was derived when the aniline value was disregarded [102]. Normally, Hammett σ values relate to a reaction on a side chain. It is therefore not expected that their use for reactions on the ring itself leads to a quantitative correlation.

More interesting are the partial rate factors

$$F_i = k_r(\text{Ø}X) \cdot \frac{P_{R_i}(\text{Ø}X)}{\sum_i P_{R_i}(\text{Ø}X)} \cdot \frac{6}{n_i}, \tag{7.2}$$

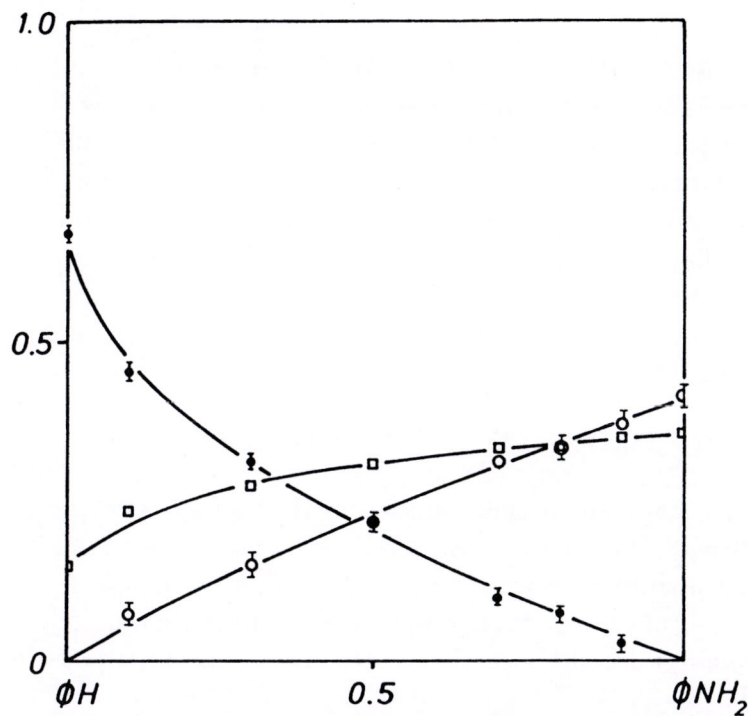

Figure 7.1: Muon polarizations observed in mixtures of aniline and benzene as a function of mole fractions. $P_D(\square)$, P_R for MuØH ($\bullet$), and $\sum P_R$ for MuØNH$_2$ ($\circ$).

Table 7.2: Partial rate factors relative to benzene for Mu and T addition to aniline.

	ortho	*meta*	*para*
Mu	1.9	1.2	1.7
T*	4.7	1.36	2.0

* Taken from ref. [66]

where n_i is the number of equivalent sites for formation of the i-th radical, i.e. 1 for the *para* and 2 for the *ortho* and *meta* positions. The factor F_i combines the information on intramolecular selectivity with that on intermolecular selectivity and relates the reactivity of one carbon atom towards Mu addition in ØX to the analogous property in ØH. The results for aniline are given in Table 7.2. They show that NH_2 has an activating effect in particular in the *ortho* and *para* positions and less pronounced in the *meta* position. Again, the lower selectivity of Mu compared with T is demonstrated.

7.2 Polysubstituted benzenes

We have seen that a substituent influences the regioselectivity of Mu addition in a specific way. We are now interested to see how this selectivity propagates in molecules with several substituents. Poly-methyl, poly-fluoro, di-methoxy and methoxy-cyano substituted benzenes were investigated [69,77]. The principal conclusions are the same in all cases, and the methyl series suffices as an example.

The polarizations of 16 radicals observed in benzene, toluene, the three xylenes, and in three trimethyl-benzenes were fitted to a relation analogous to eqn. (5.2) for the hyperfine coupling constants

$$P_R = n P_0 \Pi_X (1 + \Delta_X), \tag{7.3}$$

where n is the number of equivalent sites for Mu addition leading to the same radical, Δ_X represents a position-dependent substituent effect, and P_0 is a proportionality constant. In principle, P_0 may change from one substance to another. In particular, it is expected to decrease on substitution with electron scavengers. It was however found to be essentially constant for the cases studied and was thus taken the same for all molecules. The calculated values are plotted against the observed ones in Figure 7.2. An excellent

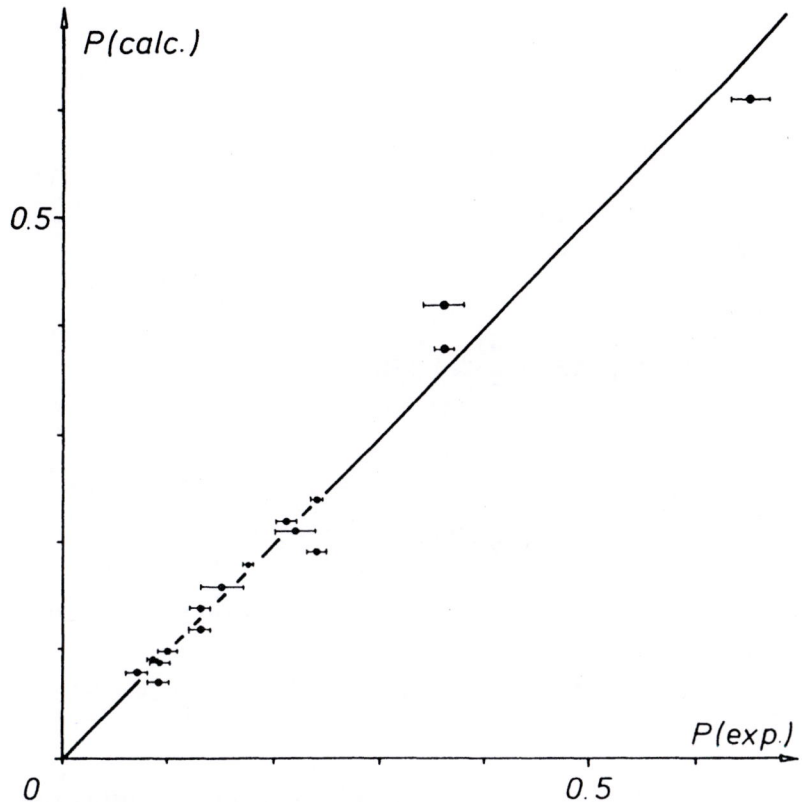

Figure 7.2: Correlation between experimental and calculated product distributions of Mu addition to poly-methyl substituted benzenes.

correlation is obtained. The best-fit values are $P_0 = 0.101(2)$, $\Delta_o = +0.17(2)$, $\Delta_m = -0.11(2)$, and $\Delta_p = -0.09(2)$. This demonstrates again that CH_3 has an *ortho*-directing effect to Mu addition at the cost of the *meta* and *para* positions. It is gratifying that a cumulative relation (7.3) holds since it allows the prediction of product distributions in polysubstituted benzenes based on the observations in the monosubstituted case, e.g. for Mu addition from the data in Table 7.1.

Chapter 8

Radical reactions

8.1 Secondary isotope effects

A reaction is said to involve a secondary kinetic isotope effect when no bond to the isotopic atom is broken or formed in the rate-determining step. A significant effect can be expected only when there is some force constant change between reactant and transition state for normal modes which involve the isotope. This can arise from an alteration in the type of bonding, such as a change in hybridisation, or it can be a change in non-bonding interactions of the isotopic atom, as in steric effects [19].

In muonated free radicals the isotope is bound in the β-position to the reactive centre, or more remotely if the unpaired electron is delocalized. β-deuterium kinetic isotope effects have been observed mainly for S_N1 solvolytic reactions, where they run about 5-15% per deuterium atom [19]. The effect is ascribed to isotope dependent hyperconjugative stabilization of the product carbonium ion, which seems to be more efficient for $C-H$ as compared with $C-D$ bonds. Such effects are less important in free radical processes where the p-orbital of the reactive centre is singly occupied and thus less electron deficient than in reactions involving carbonium ions. k_H/k_D is usually found to be 1.02 per deuterium atom, but it was 1.08 in a reaction which was thought to be the most favourable for a large kinetic isotope effect [19].

It is difficult to find experimental evidence for effects by comparison of the reactions of muonated radicals with those of their H analogues since they have almost never been measured under exactly comparable conditions. Furthermore, conventional radical rate constants are often measured relative to those of 'clock' reactions, which limits their use for accurate absolute comparisons. A better idea is obtained from reversible reactions where Mu induces an energy difference between otherwise degenerate reactant and product states, as for example in the inversion of the cyclopentyl radical. There, the conformer with Mu in a more axial position is lower in energy by *ca.* 1.4 $kJ \cdot mol^{-1}$ than its inverted conformer. Assuming equal frequency factors one concludes that the forward and backward reactions both exhibit an isotope effect of *ca.* 30%. This is a case with a clear conformational change. In other reactions the effects may be smaller.

8.2   Electron transfer reactions

Substituted cyclohexadienyl type radicals react with oxidants (Ox) by electron transfer [103,104]:

$$\hspace{11cm} (8.1)$$

In studies employing radiolytic radical production in aqueous solution it was found that the rate constants for oxidation decrease and the selectivity of isomeric cyclohexadienyl radicals increases as the redox potential of the oxidant decreases. However, it is often difficult to determine the distribution of isomeric radicals since the isomers differ only slightly with respect to their physical and chemical properties. Rate constants have to be deduced by extraction of different components in the build-up kinetics of Ox⁻ in combination with product analysis. This requires a profound knowledge of all species formed in the radiolysis and of their reactions, which may complicate the analysis considerably.

The reaction of the unsubstituted muonated cyclohexadienyl radical with benzoquinone (BQ) was the first example to demonstrate that rate constants for radical reactions can be determined [49]. Since then, the system was extended [105] to include duroquinone (DQ) as oxidant, and methyl and methoxy substituted cyclohexadienyl radicals.

Chemical reaction leads to lifetime broadening of the radical lines (eqn. 2.6). This is demonstrated in Figure 8.1 which shows one of the cyclohexadienyl lines in pure benzene and with increasing concentrations of DQ. The solid line is the best fit of the theoretical expression to the experimental points in Fourier space. The line width parameter λ is plotted *vs.* concentration of the oxidant in Figure 8.2 for C_6H_6Mu and C_6D_6Mu. It demonstrates nicely the linear behaviour expected from eqn. 2.7. The intersept is larger for C_6H_6Mu because of the onset of splitting which was demonstrated in

73

Figure 4.1. Within the experimental error of 15% the slopes of the two lines are the same. This absence of a kinetic isotope effect between H and D substituted radicals suggests that the effect between Mu and H is also not too severe. Actually, the product of the reaction is a cation. In line with the discussion in Section 8.1 one expects that the stabilizing effect is greater for the lighter isotope. If a kinetic isotope effect should be present it should be expected to favour the reaction of the radical with the lighter isotope.

The selectivity of the method is demonstrated in Figure 8.3. The three isomers are well resolved in the spectrum obtained with pure anisole. Upon addition of 2.5 mM BQ the lines corresponding to the *para* isomer disappear in the noise since they are too broad at this concentration. Furthermore, the apparent *ortho/meta* intensity ratio changes as a consequence of the higher rate constant for the *ortho* isomer.

The rate constants measured for several of these reactions are collated in Table 8.1. k_{BQ} is 4-9 times larger than k_{DQ}. This reflects the higher oxidizing power of BQ and is in accord with expectation. Furthermore, the rate constant for the *meta* substituted isomers are quite the same as those for the unsubstituted radicals. This is reasonable for an oxidation reaction since the electron to be transferred occupies a molecular orbital with a node in the *meta* position. For the *ortho* and *meta* methyl substituted isomers the rate constants are nearly the same, showing that CH_3 is a weak substituent. Qualitatively, this is also true for the *para* isomer, but accurate numbers were not obtained since the corresponding weak lines overlapped with the stronger lines of the *ortho* isomer at higher solute concentrations. The OCH_3 substituent appears to have a significant accelerating effect in the *ortho* and in particular in the *para* position. This was observed also for the oxidation of methoxyhydroxy-cyclohexadienyl radicals and was ascribed to the solvation of contributing polar structures in the transition state [103]. There, the effect was greater, which must be a consequence of the higher polarity of the solvent, H_2O. In this context it is noted that k_{DQ} for C_6H_6Mu is $9 \cdot 10^7 M^{-1}s^{-1}$ and thus increased by 50% when a 1:1 mixture by weight of benzene and methanol is used as a solvent instead of pure benzene. This reflects the expected acceleration due to stabilization of the ionic reaction products and in particular of the polar transition state. All rate constants are clearly below the diffusion controlled limit, but preliminary experiments reveal an activation energy of only about 2.3 kJ·mol^{-1} for k_{BQ} of the methoxy substituted radicals [105]. This suggests that the reaction mechanism may be more complex than given in eqn. (8.1).

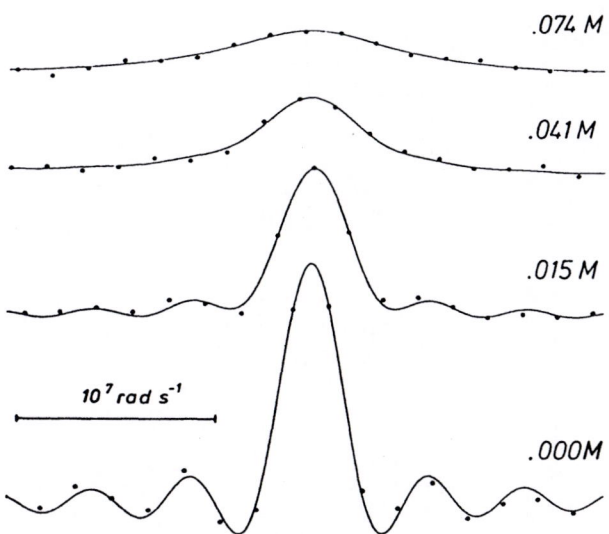

Figure 8.1: Line width effect on a C_6H_6Mu radical frequency in solutions of duroquinone in benzene.

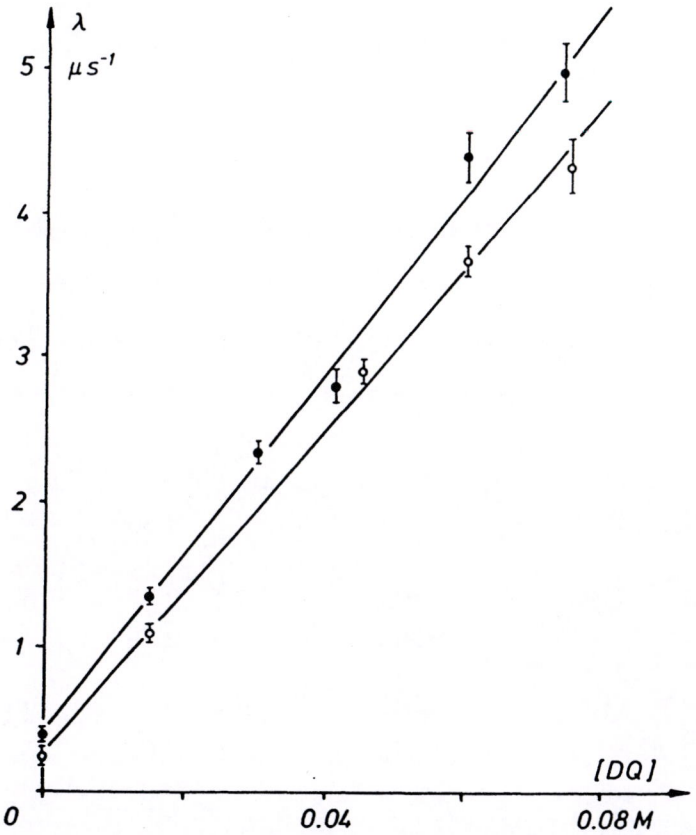

Figure 8.2: Line width parameter λ for C_6H_6Mu ($\bullet$) and for C_6D_6Mu ($\circ$) as a function of duroquinone concentration.

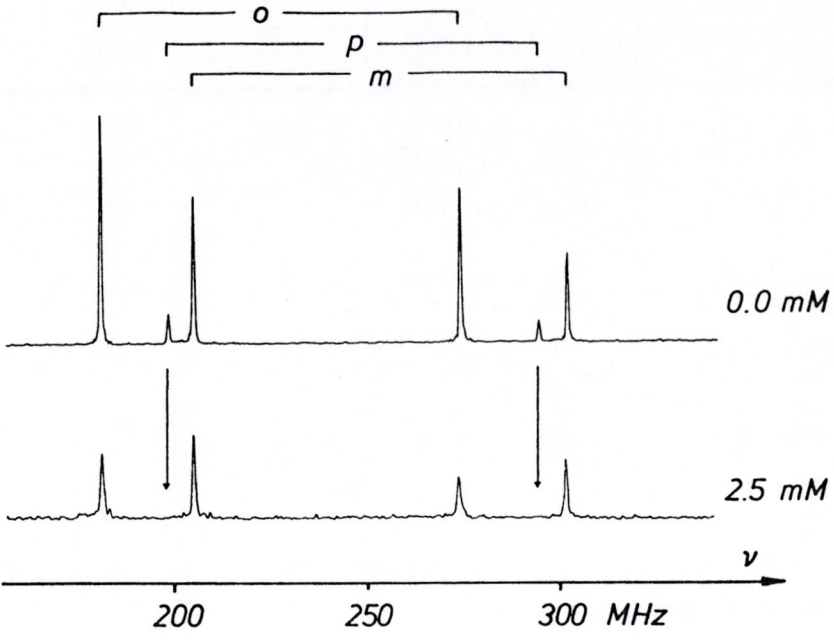

Figure 8.3: Fourier power μSR spectra obtained with neat anisole and with 2.5 mM BQ in anisole.

Table 8.1: Rate constants for the reaction of cyclohexadienyl radicals with quinones at 293 K

solvent	substituent	isomer	$10^{-8}k_{BQ}$ $M^{-1}s^{-1}$	$10^{-8}k_{DQ}$ $M^{-1}s^{-1}$	k_{BQ}/k_{DQ}
benzene	H		2.6	0.62	4.2
benzene-d_6	D[a)]		—	0.56	—
toluene	CH$_3$	ortho	3.4	0.52	6.5
		meta	3.5	0.79	4.4
anisole	OCH$_3$	ortho	10.5	1.2	8.8
		meta	4.2	0.66	6.4
		para	$\approx$20	—	—

[a)] The radical is C$_6$D$_6$Mu

76

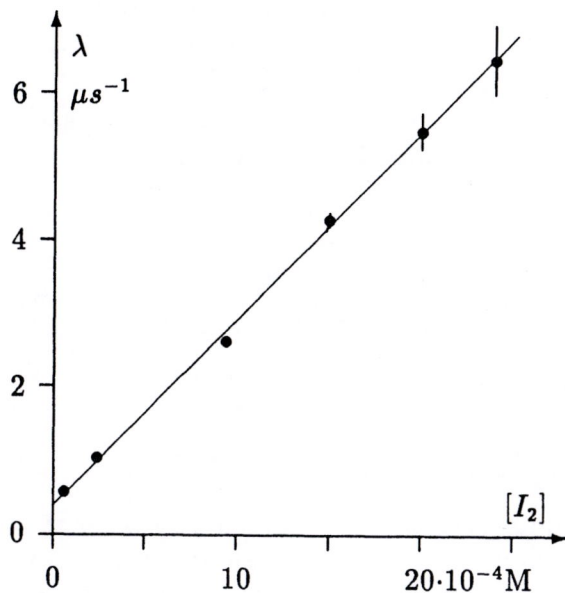

Figure 8.4: Reaction of the cyclohexadienyl radical with I_2.

8.3 Abstraction reactions

Iodine is commonly used as an efficient radical scavenger in hot atom chemistry. In μSR experiments of dilute solutions in benzene the width of cyclohexadienyl radical lines increases linearly with iodine concentration (Figure 8.4). The rate constant deduced is $k = 2.5 \cdot 10^9 M^{-1} s^{-1}$. This value is close to the limit of diffusion control. It is ascribed to the abstraction reaction

$$C_6H_6Mu + I_2 \longrightarrow C_6H_6MuI + I. \tag{8.2}$$

3 M methyliodide does not lead to noticeable line broadening, which indicates a rate constant which is at least four orders of magnitude lower than for I_2. This is understood in terms of the C–I bond dissociation energy, which is 236 kJ·mol^{-1} for ICH$_3$ but only 153 kJ·mol^{-1} for I_2.

8.4 Spin exchange reactions

Radicals are electron spin doublet species ($S = \frac{1}{2}$). A pair of two interacting radicals is usually described in the basis of its zero-field spin functions with total electron spin quantum number $S = 0$ (singlet state, S), or $S = 1$ (triplet state, T). The latter decays into the three substates T_+, T_0, and T_-, with corresponding quantum numbers for the z-component, $M_s = +1$, 0, or -1. For an encounter pair of spin-uncorrelated radicals (A,B) each of the four states has the same weight. Under conditions of spin conservation, strong spin exchange (exchange frequency large compared with inverse encounter duration) and diffusion controlled chemical reaction this leads to the following situation.

isolated reactants	combined spin state		products	chemical reaction	spin exchange
$\uparrow^A\uparrow^B$	T_+	$\rightarrow$	$\uparrow^A\uparrow^B$	0	unnoticeable
$\uparrow^A\downarrow^B$, or $\downarrow^A\uparrow^B$	T_0	$\rightarrow$	$\uparrow^A\downarrow^B$, or $\downarrow^A\uparrow^B$	0	50%
$\downarrow^A\downarrow^B$	T_-	$\rightarrow$	$\downarrow^A\downarrow^B$	0	unnoticeable
$\uparrow^A\downarrow^B$, or $\downarrow^A\uparrow^B$	S	$\rightarrow$	A-B, or C + D	100%	none

S pairs terminate chemically by combination to A-B or by disproportionation to C + D. T pairs can usually not terminate since the T product states are too high in energy. Only for half of the T_0 states spin exchange leads to products with opposite spins. Thus, 25% of all encounter pairs react chemically and 12.5% lead to noticeable spin exchange.

The partition between spin exchange and chemical termination is a sensitive function of reaction conditions. Activated termination slows down the chemical channel, but half of the separating S pairs will show the effect of spin exchange. Weak exchange interaction of course decreases the exchange efficiency without affecting termination. Determination of the rate constant and of the partition is thus a sensitive test for the interaction of two radicals.

The system investigated was the stable free radical 2,2-diphenyl-1-picryl hydrazyl (DPPH) in benzene. The rate constant of encounter with a cyclohexadienyl radical was calculated from

$$k_D = 4\pi \cdot N_L \cdot 10^3 (R_A + R_B) \cdot (D_A + D_B). \qquad (8.3)$$

Using encounter radii of 0.274 nm and 0.419 nm derived from group increments for benzene and DPPH, and diffusion constants of $2.15 \cdot 10^{-9} \mathrm{m^2 s^{-1}}$ and $0.76 \cdot 10^{-9} \mathrm{m^2 s^{-1}}$ measured by the capillary broadening technique for 1,2-cyclohexadiene and 1,1-diphenyl-2-picryl hydrazine (the hydrated form of DPPH) in benzene at 298 K we obtain $k_D = 1.53 \cdot 10^{10} \mathrm{M^{-1} s^{-1}}$. Assuming strong exchange and diffusion controlled termination we calculate a total rate constant of $5.74 \cdot 10^9 \mathrm{M^{-1} s^{-1}}$, $\frac{2}{3}$ of it being chemical termination and $\frac{1}{3}$ spin exchange.

An electron spin flip swaps the two radical lines in high transverse field μSR and has therefore the same effect on the line width as chemical reaction.

The experimentally determined rate constant of $5.5(2) \cdot 10^9 M^{-1}s^{-1}$ reflects the sum of both channels. It is barely below the predicted number, which demonstrates that termination is close to diffusion controlled and exchange is strong.

The situation is different in longitudinal fields. Chemical reaction leads to simple broadening of the Lorentzian ALC resonance (eqn. 3.29). Broadening due to spin exchange is a different function of the rate, so that a combination of transverse and longitudinal experiments should allow separation of the two processes. Furthermore, for a sufficiently high transition frequency the signal amplitude is predicted to pass through a maximum since muons not previously in resonance start to precess after electron spin flip [43]. The latter effect was verified experimentally. Figure 8.5 shows the 2.08 T line observed with pure benzene and with two concentrations of DPPH. The signal amplitude at intermediate concentration is clearly the highest, but the preliminary nature of the data precludes more quantitative statements about the partition between the two processes.

8.5 Transfer of bound Mu, an example for a primary kinetic isotope effect

Most studies of kinetic hydrogen isotope effects involve reactions where a reagent attacks a H–R/D–R bond, i.e. the isotope is the atom to be transferred, whereas in studies of Mu reactions the isotope is the attacking species. There is however one example of the former type [106]. It deals with Mu transfer from cyclohexadienyl radicals to 2,3-dimethyl-1,3-butadiene (DMBD).

One radical is observed in each, pure benzene and pure DMBD, and they both show narrow lines ($\lambda_0 < 0.5\mu s^{-1}$). In binary mixtures of the two, and in three-component mixtures with cyclohexane, the cyclohexadienyl lines broaden with increasing concentration of DMBD (Figure 8.6) whereas the line width of the allyl type radical remains unchanged. This is obviously due to a reaction of the cyclohexadienyl radical with DMBD. One obtains a rate constant of $k_{Mu} = 0.95 \cdot 10^6 M^{-1}s^{-1}$ at 293 K, and $\log(A/M^{-1}s^{-1})=6.9(4)$ and $E_a = 6(3)$ kJ $\cdot$ mol^{-1} for the Arrhenius parameters. In time resolved ESR experiments the protiated cyclohexadienyl radical was observed, but it did not react significantly with DMBD. The second order rate constant, $k_H \cong 12$ M^{-1}s^{-1}, has to be regarded as an upper limit. Comparison of the two values reveals a huge kinetic isotope effect, $k_{Mu}/k_H \gtrsim 7.5 \cdot 10^4$ at room temperature. This seems possible only when Mu is directly involved in the reaction. It is believed that the reaction is

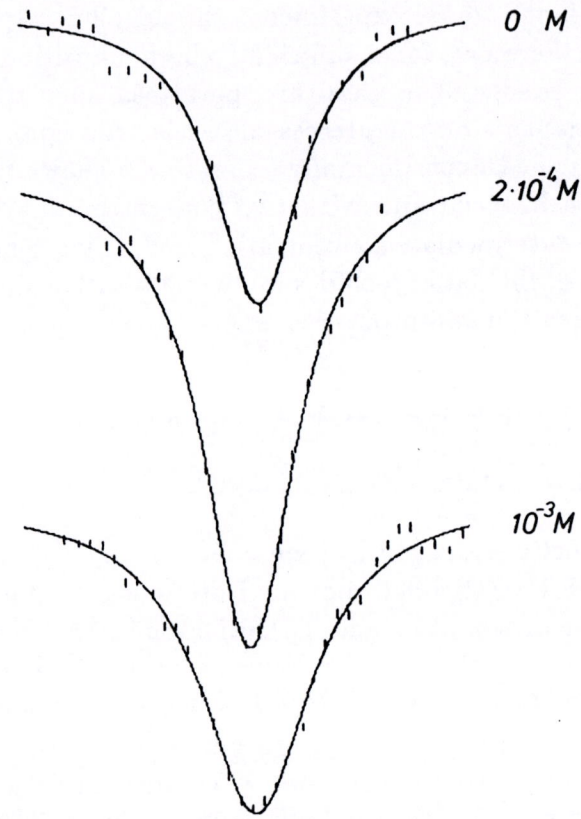

Figure 8.5: ALC resonance at 2.08 T observed with pure benzene and with solutions of DPPH.

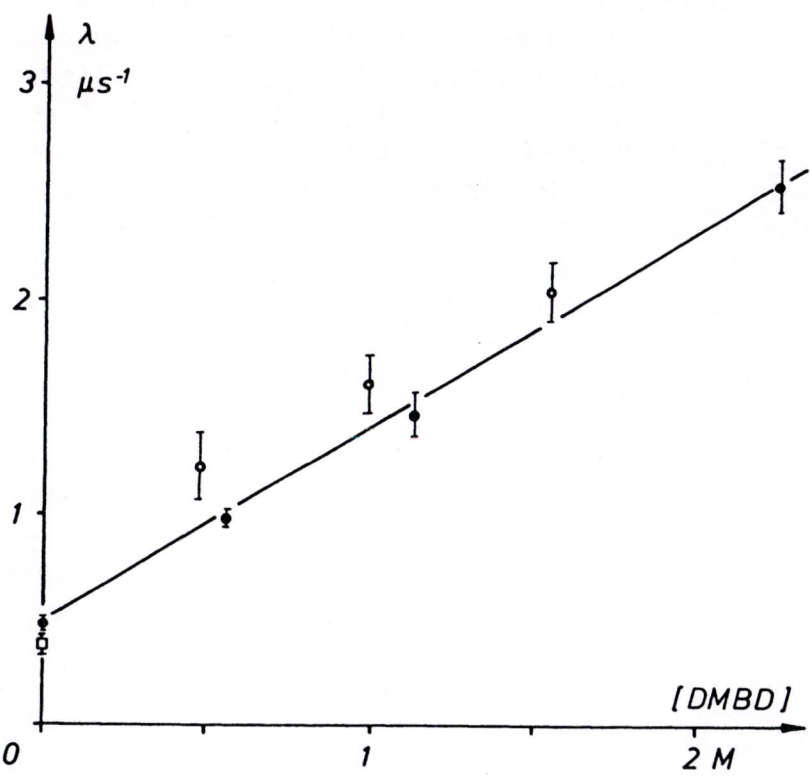

Figure 8.6: Width of C_6H_6Mu lines in solutions of various compositions at 293 K [106].

$$\text{(8.4)}$$

since it is exothermic by about 80 kJ·mol^{-1} due to the recovery of delocalization energy in benzene. Reaction of C_6H_6Mu is highly favoured over that of C_6H_7 due to its higher zero-point vibrational energy for the motions of the lighter isotope. A rough estimate considering only the stretching mode with $\tilde{\nu} \approx 3000$ cm^{-1} for C−H and correspondingly 8680 cm^{-1} for C−Mu yields $\Delta H \approx 34$ kJ·mol^{-1} between C_6H_7 and C_6H_6Mu. This alone would lead to $k_{Mu}/k_H \cong 10^6$. The factor is reduced if zero-point energy in the transition state is taken into account, but enhanced if all normal modes are considered, and in particular if we allow for tunneling of the low mass Mu through the lower and therefore also narrower barrier. The low frequency factor for k_{Mu} indeed indicates a large contribution of tunnelling. Kinetic isotope effects of several orders of magnitude along with convincing evidence for tunnelling have been obtained for H/D transfer reactions at low temperatures, e.g. for the isomerization of sterically hindered aryl radicals [107], but at room temperature a value as large as the one reported for reaction (8.4) has not been attained previously.

Chapter 9

Summary and review

9.1 Objective

Throughout this work, we have used cyclohexadienyl type radicals to exemplify how the μSR techniques work, and what their potential and limitations are. We now summarize these results and supplement them with observations made with other types of radicals. We shall see that the problems encountered are similar in most cases, but some nice new features are worth mentioning.

9.2 The requirements for the observation of muonated radicals

Muonated cyclohexadienyl radicals have been derived from corresponding aromatic parent molecules, formally by Mu addition. Likewise, all other muonated radicals observed today are derived from unsaturated parent compounds. None of them was obtained with saturated molecules.

There is no way to alter the preferred direction of Mu addition to a given compound. Thus, primary radicals for example are hard to produce since terminal addition to an olefin is always preferred over non-terminal addition.

In transverse fields the phase coherence of the muon precession in the radical is lost when the Mu precursor lifetime exceeds $\approx 10^{-10}$s. This can impose a severe limitation for the observability of a radical. Addition to certain compounds is activated, and this is the most probable reason for the non-observation of the Mu adduct to acetonitrile and to P=O and S=O double bonds. In solutions, the concentration of the unsaturated solute must be $\gtrsim 0.05$ M even for diffusion controlled addition. Here, the ALC technique

83

is far more sensitive. It is sufficient to have appreciable radical formation in the muon lifetime. This requirement is less stringent by three orders of magnitude in comparison with transversie field μSR. It should give access to the observation of a whole new class of radicals. In particular, it should also become possible to observe paramagnetic products of chemical reactions. This eliminates one of the most severe limitations of the conventional techniques. So far, only in one case a diamagnetic product of a slow reaction was observed with an alternative longitudinal field technique which uses an oscillating field B_1 perpendicular to B_0, in analogy to CW magnetic resonance [30].

In cases of internal dynamics, radicals may be observed in their low temperature equilibrium conformations (slow exchange limit) as long as they have a lifetime of $\gtrsim 0.2$ μs. On the other hand, an average structure may be observed in the fast exchange limit. For a two-sided degenerate exchange the transverse field line width is given by [108]

$$\lambda = \lambda_0 + \frac{(A_\mu^I - A_\mu^{II})^2}{4 \cdot A} \cdot e^{+E_a/RT}. \tag{9.1}$$

A is the Arrhenius frequency factor which is normally of the order $10^{13}\mathrm{s}^{-1}$. Assuming a difference of 100 MHz between the two coupling constants the radical thus becomes observable at room temperature for an activation energy $E_a \lesssim 13$ kJ·mol^{-1}. Most of the vinyl type radicals and some cyclic alkyl radicals are not observed because they are in the intermediate exchange region near room temperature [70]. In ALC experiments, radical signals disappear in the same temperature range on the slow exchange side. The positions of the resonances is given by the difference between muon and proton coupling constants (eqn. 3.24). Therefore, the lines may reappear on the fast exchange side in a different temperature range than in transverse fields.

9.3 The different types of radicals observed in liquids

At least 250 different muonated radicals have been observed so far, many of them in the first years when it was important to study the chemical principles of the formation of muonated radicals and the relations between structure and hyperfine coupling constants. With roughly 120 members the cyclohexadienyl radicals represent the largest family. About 70 of them are monosubstituted derivatives, more than has been observed by ESR [23]. The easy method of producing them and the simple two-line spectra render μSR superior to other techniques for this group of radicals.

Primary alkyl type radicals are generated from terminal olefins. They are generally observed as weak features since they are formed in competition with the secondary species, which are favoured on energetic grounds. Unbranched nonterminal and cyclic olefins lead to secondary radicals. The formation of tertiary radicals is favoured over that of secondary species where both possibilities exist. About 50 muonated alkyl type radicals containing no elements other than C, H, and O have been reported [34,70,109-113]. Furthermore, 17 allyl type radicals derived from conjugated dienes have been observed [110] and the regioselectivity of their formation studied [89]. Other strongly conjugated radicals were observed with styrene and with acrylonitrile [110].

7 radicals were obtained with a series of chloro-olefins but many more were expected to be formed [114]. The latter may escape observation because of rearrangement or elimination reactions.

Species with certain substituents X in β-position show relatively rigid conformations due to a strong hyperconjugative interaction of X with the radical centre. Muonated examples with X=Cl, SiMe$_3$, Si(OMe)$_3$, PO(OEt)$_2$, SO$_2$, SO$_3^-$, or Sn(n-Bu)$_3$ have been observed, often along with the α-substituted isomer [77,115].

The Mu adduct to the carbonyl oxygen of acetone was among the first muonated radicals reported [8]. Later on, analogous species were observed with cyclohexanone [99] and with a few other ketones, aldehydes, esters, and amides [77,116]. They all exhibit low coupling constants for the muon, and often the signals are weak. Oxygen adducts are also found with nitrobenzene [69] and with a number of nitroalkanes [77].

Radicals similar to the carbonyl Mu adducts are the adducts to the sulfur of C=S, as in CS$_2$ [32] and in thiobenzophenone [77]. In the case of CS$_2$ the reduced muon-electron hyperfine coupling constant A'_μ is only 0.8 MHz, which is the smallest value observed so far. Some other C=S-containing molecules lead to radicals with much larger coupling constants which suggest that they are due to Mu adducts on the side of the carbon atom [117].

7 muonated hydrazyl radicals were observed in a series of symmetric and asymmetric azo-compounds [118]. For these nitrogen adducts the coupling constants are of the order of those found for the carbonyl compounds. Addition to carbon-nitrogen double bonds seems to occur preferentially on the side of the carbon atom [118].

A few triple-bond Mu adducts were also observed. In phenyl acetylene this leads to a linear structure at the radical centre [119], but with SiMe$_3$-substituted acetylenes typical vinyl-type radicals are formed [120]. The disilyl derivative shows $A'_\mu = 230$ MHz, the largest value yet obtained for a muonated organic radical. An interesting case arises with 1,1-dimethylallene [121]. Addition at the methylene carbon gives the trimethyl-vinyl radical, at the central carbon it leads to the 1,1-dimethylallyl radical. The latter is the only organic case where Mu is restricted to the nodal plane of the π-system containing the unpaired electron and therefore the best approximation so far for a radical with Mu in α-position. The only comparable system is Mu* in the GaAs semiconductor. There, the muon was located recently close to the centre of a Ga−As bond [122]. It occupies a position near a node of the wave function. The spin populates mainly the two neighbouring atoms, as in allyl, and the sign of the hyperfine coupling constant is positive on all three atoms. A closer approximation to an α-position may be found for Mu* in diamond, silicon and germanium. In all these cases the isotropic hyperfine coupling is negative [7], and for diamond it is close to what one would expect for α-Mu in the methyl radical.

9.4   Observations in other phases

Cyclohexadienyl was the first radical observed in a polycrystalline solid [49]. Tetramethyl cyclohexadienyl was the first radical detected in a single crystal (Section 4.2.1), and it was again cyclohexadienyl which was first observed adsorbed on a surface (Section 4.2.2). A more detailed study of a surface adsorbed species was obtained with the allyl type radical derived from dimethylbutadiene [54]. Information about the anisotropic reorientational motion on the surface was derived, and in polycrystalline material a powder pattern was obtained which showed all three components of the hyperfine tensor. Another example which nicely demonstrates anisotropic motion was obtained for the radical derived from norbornene in a plastic crystalline state [123]. The hyperfine anisotropy increased with decreasing temperature, which indicates a tendency of the radicals to align more and more along a preferred orientation as temperature is reduced.

An interesting situation arises by muon irradiation of polyacetylene, an organic polymer semiconductor [124]. In its *cis*-modification, a radical with A_μ=91 MHz was observed. The large line width indicates a distribution of chain lengths and possibly of conformations near the Mu substitution site. No radical was observed in the *trans*-form. Instead, a relaxation was found in longitudinal fields and interpreted in terms of a one-dimensional soliton motion of the electron spin along the chain. The different behaviour was ascribed to the degeneracy of resonance structures for radicals derived from the *trans* but not from the *cis* modification. In the latter case, motion of the unpaired electron would require bond length alterations, as it is seen in the following structural schemes (the asterisk denotes the Mu substituted carbon atom):

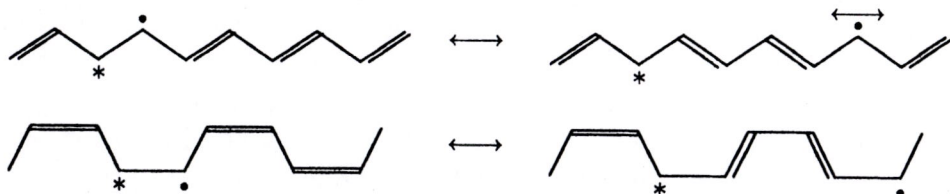

The first observation in the gas phase was that of the muonated ethyl radical by transverse field time resolved experiments at pressures between 25 and 50 atm [125]. The decreasing line width with increasing pressure was due to collisional narrowing. At lower pressures, the broad lines would render detection difficult in transverse fields. A longitudinal field ALC observation was reported for low pressure ethylene between the grains of a fine SiO_2 powder [48], but there is no clear evidence that the radical is indeed in the gas phase and not adsorbed at the grain surface.

9.5 Investigation of isotope effects

Isotope effects are often interesting *per se* since they allow the separate study of mass dependent phenomena. Mu is particularly attractive in this respect due to the extreme mass ratio compared with other hydrogen isotopes. For the isolated atom, the prime interest lies in the field of kinetics. Substituted in radicals, it leads to a pronounced isotope dependence of hyperfine interactions.

It was found that the reduced muon coupling constant is always larger than the proton coupling in the analogous hydrogen radical. This has different origins, and the extent of their contributions is still under discussion. Within the Born-Oppenheimer approximation the electronic wave functions of isotopic species are the same, and the differences have to be attributed to vibrational and other internal degrees of freedom. In an anharmonic potential the higher zero-point vibrational energy associated with normal modes involving the lighter isotope results in an increased bond length of C−Mu as compared with C−H, and this leads to a higher coupling constant. Even in a harmonic potential the region of space sampled by the muon is larger owing to the larger vibrational amplitude, and consequently the vibrationally averaged coupling constant is isotope dependent (compare Section 4.3.3).

In all these cases a pronounced preference was found for a conformation with the C−Mu bond eclipsing the half-filled p_z-orbital at the radical centre. The potentials hindering internal rotation in ethyl radicals were shown to depend markedly on isotopic substitution [111,112]. This is also explained in terms of internal dynamics. It is ascribed mostly to the difference in zero-point vibrational energies between ground and transition state of the internal rotation, as in the case of secondary isotope effects in chemical reactions [18,126]. The force constant for the C−X stretching motion is lower in the eclipsed conformation since the overlap of bond and half-filled orbital facilitates π-bond formation upon stretching and lowers the corresponding dissociation energy. An additional contribution may be of steric origin, since the van der Waals radius is much larger for the lighter isotope. Mu thus appears bulkier than H, and steric interaction with other substituents is minimized when the C−Mu bond eclipses the half-filled orbital at the radical centre.

An alternative explanation for the preferred C−Mu bond orientation invokes mass dependent hyperconjugation [127]. In the light of the BO approximation such a terminology seems delicate, and the zero-point vibrational interpretation should clearly be preferred.

A careful theoretical analysis based on non-adiabatic natural orbitals was reported for the H_2^+ family of radicals, with particular regard to the muonated members MuT^+, MuD^+, MuH^+, and Mu_2^+ [128]. It is the only quantitative discussion of the validity of the BO approximation for muonated molecules. It shows that there is no unique energy curve $E(R)$ which properly represents all of the species, but the shifts in binding energies, $E(R_e)$, are generally small. Between H_2^+ and HMu^+ it amounts to only 2.7 kJ·mol^{-1}. Along with this there is an increase of 0.65% in R_e and of 4.0% in $<R>$ for the lighter species in its lowest rovibrational state.

Interesting aspects arise from theoretical predictions of purely vibrational chemical bonding on neutral or dissociative electronic surfaces [129,130]. This should occur for example in triatomic systems

$$XHX \longrightarrow X + HX \qquad (9.2)$$

in situations where zero-point vibrational energy favours the complex XHX. This is the case for heavy atoms X (e.g. for X=I, Br), and in particular for light isotopes of H. Thus, bonding is a quantum effect arising because of the vibrationally dispersed light atom. Many aspects of it are therefore analogous in $X\cdots H\cdots X$ to those in $H^+\cdots e^-\cdots H^+$. With Mu, the lightest H isotope available, bonding should be most pronounced, and the radical nature of XMuX should allow identification of the complex, but corresponding efforts have so far remained unsuccessful.

9.6 Investigation of early events

The study of radiation chemical processes near the end of the muon track poses particular problems since they occur on a sub-nanosecond timescale and are therefore not accessible to direct observation. Indirect information is obtained from magnetic field effects provided that precursors of the observed species live sufficiently long to allow for phase shifts and for loss of polarization (Section 3.2.2). Furthermore, scavengers are able to interfere in these processes and thereby alter the distribution of muon polarization between the various chemical species.

Radical formation is too fast in pure aromatic compounds to allow for magnetic field effects. The nature of the radical precursor is therefore not known. It was however shown that Mu is the precursor in benzene/cyclohexane mixtures (Section 6.2). Furthermore, muon polarizations exhibit a typical scavenging behaviour in solutions of CCl_4 and of I_2 in benzene. This interference of electron scavengers suggests that end-of-track electrons are involved in the formation of the radical or of its Mu precursor.

In many organic liquids one observes only muons in diamagnetic environments, and there is a large fraction P_L of missing muon polarization. Some of it could constitute Mu which is short-lived on a microsecond timescale, either by virtue of reactions with unsaturated impurities or with the substrate itself. In spin trapping experiments utilizing 2,3-dimethylbutadiene (DMBD) as trapping agent it was possible to observe the allyl type radical derived from the diene. From the phases it was possible to unambiguously demonstrate that Mu is formed in various organic liquids including acetonitrile, dimethylsulfoxide, N,N-dimethylformamide, and acetone [32]. Mu was detected even in an electron scavenger like $CHCl_3$, but not in CCl_4. The latter is particularly gratifying since P_D in CCl_4 was always taken to be unity and used as a reference for absolute polarizations. Mu was also not formed in CS_2 [32]. The reason for this is not clear. It demonstrates that the conditions for charge neutralization at the end of a muon thermalization track are not yet fully understood.

Some symmetrical azo-compounds are exceptional as P_D and P_R add up to unity and the lost fraction of polarization disappears [118]. These liquids may provide a clue for a more complete understanding of the lost fraction in future.

Detailed information has been obtained for end-of-track effects and radical formation processes in acetone [44]. It was shown that the 2-propylol radical is formed via two parallel channels. One involves Mu with a lifetime of 0.1 ns and a second one with diamagnetic Mu^+ and solvated electrons with a lifetime of 3 ns as precursors (compare eqns. 6.1-6.3). The same results were obtained from magnetic field effects with pure acetone and from scavenging experiments with DMBD. The average initial separation of Mu^+ from its closest spur electron was estimated to be 3-4 nm, which is a typical secondary electron range in polar organic liquids.

9.7 Investigation of radical kinetics

Cyclohexadienyl type radicals served to demonstrate that the muon can be used as a probe to measure rate constants for radical reactions. Besides the spin exchange, electron transfer and abstraction reactions which were discussed in Chapter 8 a considerable set of other reactions have been studied. There are in particular the radical rearrangements, i.e. cyclopropyl ring fission of the type

and 1,5-cyclization reactions, such as

The asterisk denotes the muonated methyl group. The influence of substituents on frequency factors and activation energies were interpreted in terms of polar and steric effects [34,109].

The isomerization kinetics of partly delocalized radicals of the type

(R_1=H or CH_3, R_2=CH_3, OC_2H_5 or $N(CH_3)_2$) was also studied [113]. Most of them showed normal Arrhenius parameters with $log(A/s^{-1}) \approx 13.2$ and $E_a \approx$ 45 kJ $\cdot$ mol^{-1}. Exceptionally low values for the 1-methylacetonyl radical agree with those derived from ESR measurements, but they are not understood.

Further reactions include the addition of the *tert*-butyl radical to acrylonitrile [131] and of the 2-propylol radical to 2,3-dimethylbutadiene [44]. For the former, the rate constants determined by modulated ESR and by μSR were $10^6 M^{-1} s^{-1}$ and $2.4 \cdot 10^6 M^{-1} s^{-1}$, respectively. The force constants for the C−Mu stretching and bending modes increase slightly in going from the *tert*-butyl to the isobutane-like environment of the adduct [132]. Therefore, based on the theory of secondary isotope effects, one should have expected a slightly lower rather than a higher rate constant for the muonated radical. The ESR measurements were performed near the limit of the technique, but this is not expected to cause a difference of more than a factor of two. The discrepancy therefore remains unexplained. It should be noted in this context that the above cyclization reactions are intramolecular additions and as such directly comparable to *tert*-butyl addition to acrylonitrile. The environment of Mu undergoes similar changes in the course of these two reactions. Where comparison was possible, the rate constants derived from μSR agreed within the errors with those obtained with conventional methods, with a trend to lower values for μSR, as expected when secondary isotope effects play a role [18].

The inversion kinetics of the radicals derived by Mu addition to cycloheptatriene shows that the distribution pattern of the double bonds along the ring strongly influences the radical structure as well as its flexibility [108]. The observation of cyclopentyl and cycloheptyl in the fast exchange region at room temperature and the non-observation of cyclohexyl and of cyclooctyl under the same conditions demonstrates that cyclic radicals with an odd number of carbon atoms are more flexible than the others [70].

The reaction of the 2-propylol radical with CCl_4 is probably in the first step an electron transfer. The value of $1.0 \cdot 10^8 M^{-1} s^{-1}$ [96] for the rate constant agrees with the literature value for the reaction in aqueous solution [133].

The muonated radical derived from phenylacetylene showed temperature dependent line widths which correspond to Arrhenius parameters of $log(A/s^{-1}) \approx 10.5$ and $E_a \approx 12$ kJ $\cdot$ mol^{-1} [96] but it is not known what the corresponding relaxation process or chemical reaction is.

9.8 Conclusion

The extent of information which becomes available by irradiation of *ca.* 10 g of a sample with 10^8 muons is remarkable. Considering that the total amount of muonated radicals formed corresponds to only 10^{-16} mols, it is overwhelming to a chemist who is used to have a yield sufficiently large to be put on a balance. Certainly, μSR is not a synthetic method. It rather carries aspects of single photon counting techniques, except that the detected quanta are muons and positrons instead of mass-less photons. The fact that in its transverse field mode the technique permits no more than one muon at a time in the sample illustrates its extraordinary sensitivity. This is largely due to the high degree of polarization of the muon beam, which is far from the Boltzmann factors limiting sensitivity in conventional magnetic resonance experiments at temperatures where chemical reactions are normally permitted. It has the important consequences that a measurement does only negligible damage to the sample, and that kinetics of bimolecular reactions is of perfect pseudo-first order since there is no substrate turnover.

The chemical nature of muons in diamagnetic environments is still a mystery, and due to the lack of resolution of the method it will be difficult to study also in future. On the other hand, it is one of the strengths of the technique that the various radicals are ideally resolved in the spectra, and the superpositions of spectra are separated without any problems. Even the assignment of radical structures is quite trivial in most cases since, after having taken proper account of isotope effects, the muon hyperfine coupling constants behave exactly like proton couplings of the corresponding H analogous radicals, and these are well-known from three decades of ESR experiments. Moreover, the structures of the radicals formed are quite predictable so that little new information emerges from pure spectroscopic studies.

There are still many unknowns in the description of the muon's end-of-track processes which govern its distribution between different chemical species. The situation is probably more complex than the models presented here may suggest. However, a number of major processes have been traced down. The case of CS_2 shows that these may vary considerably from one medium to another, and it is encouraging to see how well such situations are distinguished in careful experiments.

The most important applications of the technique are clearly in the field of dynamical processes. Thereby, the timescale is set by the muon's microsecond lifetime. For radical kinetics, this implies a restriction to very fast reactions, but they are exactly the ones which are difficult to access by conventional means. In literature, rate constants are often reported relative to other reactions. μSR yields absolute values, and the errors obtained for rate constants and for Arrhenius parameters have been shown to be competitive with those of more conventional techniques. It can thus be used to calibrate *'clock reactions'* which serve as standards for relative determinations. Eventual isotope effectsshould be taken into account. They are small as long as the force constants governing the muon's vibrational motion do not change significantly in the course of the reaction. It is no big problem in most cases, but this has to be verified carefully for each type of reaction. Isotope effects are huge as soon as the bond to Mu is formed or broken, as it was seen in the case of Mu transfer (Section 8.5).

Of high interest for further use are the applications to anisotropic reorientational averaging, as they have been demonstrated for the cases of surface adsorbed radicals and of near-globular species in plastic crystals. Similar situations may occur in other orienting systems, as for example in liquid crystals, or in clathrates.

The invention of avoided level crossing spectroscopy has considerably extended the range of applications of the muon as a probe in matter. More information on the systems becomes available, and it is to a large degree complementary to that obtained in transverse field experiments. The potential of the new technique has only started to be explored. The results are highly promising, and the impact of this second generation of μSR studies is awaited with great interest.

Bibliography

[1] J. Chappert and R.I. Grynszpan, editors, *Muons and Pions in Materials Research*, (Elsevier, Amsterdam) 1984.

[2] D.C. Walker, *Muon and Muonium Chemistry*, (Cambridge University Press, Cambridge) 1983.

[3] B.C. Webster, *Annual Reports C*, Royal Society of Chemistry, London, 1984.

[4] A. Schenck, *Muon Spin Rotation Spectroscopy: Principles and Applications in Solid State Physics*, (Hilger, London) 1985.

[5] D.C. Walker, *J. Phys. Chem.*, **85** (1981) 3960.

[6] O. Hartmann, E. Karlsson, B. Lindgren, and R. Wäppling, editors, *Proceedings of the 4th International Conference on Muon Spin Rotation, Relaxation and Resonance*, Hyperfine Interactions, Vol. **31,32**, 1986.

[7] S.F.J. Cox *J. Phys. C: Solid State Phys.*, **20** (1987) 3187.

[8] E. Roduner, P.W. Percival, D.G. Fleming, J. Hochmann, and H. Fischer, *Chem. Phys. Letters*, **57** (1978) 37.

[9] S.H. Neddermeyer and C.D. Anderson, *Phys. Rev.*, **54** (1938) 88.

[10] J.I. Friedman and V.L. Telegdi, *Phys. Rev.*, **105** (1957) 1415.

[11] A.M. Brodskii, *Zh. Exp. Teor. Fiz.*, **44** (1963) 1612 [English transl. Soviet Phys. JETP **17** (1963) 1085].

[12] J.H. Brewer, K.M. Crowe, F.N. Gygax, R.F. Johnson, D.G. Fleming, and A. Schenck, *Phys. Rev.*, **A9** (1974) 495.

[13] D.G. Fleming, M. Senba, D.J. Arsenau, I.D. Reid, and D.M. Garner, *Can. J. Chem.*, **64** (1986) 57.

[14] R.L. Garwin, L.M. Lederman, and M. Weinrich, *Phys. Rev.*, **105** (1957) 1415.

[15] P.W. Percival, E. Roduner, H. Fischer, M. Camani, F.N. Gygax, and A. Schenck, *Chem. Phys. Letters*, **47** (1977) 11.

[16] E. Roduner, P.W. Percival, H. Fischer, M. Camani, F.N. Gygax, and A. Schenck, In *Proc. Int. Symp. on Meson Chemistry and Mesomolecular Processes in Matter*, (Dubna, 1977).

[17] P.W. Percival, E. Roduner, and H. Fischer, *Adv. Chem. Series*, **175** (1979) 335.

[18] E. Roduner *Progress in Reaction Kinetics*, **14** (1986) 1.

[19] L. Melander and W.H. Saunders, *Reaction Rates of Isotopic Molecules*, (Wiley Interscience, New York) 1980.

[20] E. Roduner and H. Fischer, *Hyperfine Interactions*, 6(1979)413.

[21] I.D. Reid, D.M. Garner, l.Y. Lee, M. Senba, D.J. Arsenau, and D.G. Fleming, *J. Chem. Phys.*, **86** (1987) 5578.

[22] B.C. Garrett, R. Steckler, and D.G. Truhlar, *Hyperfine Interactions*, **32** (1986) 779.

[23] H. Fischer and K.H. Hellwege, editors, *Landolt-Börnstein, Numerical Data and Functional Relationships, New Series, Group II*, Vol. 9b (1977) and Vol. 17b (1987), (Springer, Berlin).

[24] R.W. Fessenden and R.H. Schuler, *J. Chem. Phys.*, **43** (1964) 2704.

[25] R.W. Fessenden, *J. Phys. Chem.*, **71** (1967) 74.

[26] E. Roduner and H. Fischer, *Chem. Phys.*, **54** (1981) 261.

[27] A. Abragam, *C.R. Acad. Sc. Paris, 299, Series II*, **3** (1984) 559.

[28] R.F. Kiefl, S. Kreitzman, M. Celio, R. Keitl, G.M. Luke, J.H. Brewer, D.R. Noakes, P.W. Percival, T. Matsuzaki, and K. Nishiyama, *Phys. Rev.*, **A34** (1986) 681.

[29] M. Heming, E. Roduner, B.D. Patterson, W. Odermatt, J. Schneider, H. Baumeler, H. Keller, and I.M. Savić, *Chem. Phys. Letters*, **128** (1986) 100.

[30] K. Nishiyama, T. Azuma, K. Ishida, T. Matsuzaki, J. Imazato, T. Yamazaki, and K. Nagamine, *Hyperfine Interactions*, **32** (1986) 887.

[31] Y. Miyake, Y. Ito, Y. Tabata, and D.C. Walker, *Hyperfine Interactions*, **32** (1986) 825.

[32] E. Roduner, *Hyperfine Interactions*, **32** (1986) 741.

[33] E. Roduner, G.A. Brinkman, and P.W.F. Louwrier, *Chem. Phys.*, **73** (1982) 117.

[34] P. Burkhard, E. Roduner, J. Hochmann, and H. Fischer, *J. Phys. Chem.*, **88** (1984) 773.

[35] V.G. Nosov and I.V. Yakovleva, *Zh. Eksp. Teor. Fiz.*, **43** (1962) 1750 [English transl. Soviet Phys. JETP 16 (1963) 1236.

[36] I.G. Ivanter and V.P. Smilga, *Zh. Eksp. Teor. Fiz.*, **54** (1968) 599 [English transl. Soviet Phys. JETP 27 (1968) 301.

[37] J.H. Brewer, F.N. Gygax, and D.G. Fleming, *Phys. Rev.*, **A8** (1973) 77.

[38] P.W. Percival and H. Fischer, *Chem. Phys.*, **16** (1976) 89.

[39] E. Roduner and H. Fischer, *Chem. Phys. Letters*, **65** (1979) 582.

[40] E. Roduner, In J. Chappert and R.I. Grynszpan, editors, *Muons and Pions in Materials Research*, page 209, (Elsevier, Amsterdam, 1984).

[41] T.G. Eck, L.L Foldy, and H. Wieder, *Phys. Rev. Letters*, **10** (1963) 239.

[42] D.T. Edmonds, *Phys. Rept.*, **29C** (1977) 233.

[43] M. Heming, E. Roduner, and B.D. Patterson, *Hyperfine Interactions*, **32** (1986) 727.

[44] E. Roduner, *Radiat. Phys. Chem.*, **28** (1986) 75.

[45] H. Fischer, *Kolloid-Z.*, **180** (1962) 64.

[46] M.B. Yim and D.E. Wood, *J. Amer. Chem. Soc.*, **97** (1975) 1004.

[47] E. Roduner, In K. Crowe, J. Duclos, and G. Fiorentini, editors, *Exotic Atoms '79*, page 379, Plenum Publishing Corp., 1980.

[48] P.W. Percival, R.F. Kiefl, S.R. Kreitzman, D.M. Garner, S.F.J. Cox, G.M. Luke, J.H. Brewer, K. Nishiyama, and K. Venkateswaran, *Chem. Phys. Letters*, **133** (1987) 465.

[49] E. Roduner, *Hyperfine Interactions*, **8** (1981) 561.

[50] E. Roduner, *Chem. Phys. Letters*, **81** (1981) 191.

[51] L. Kevan and L.D. Kispert, in: *Electron Spin Double Resonance Spectroscopy*, (Wiley, New York) 1980.

[52] H. Pfeiffer, *Phys. Rep. C*, **26** (1974) 294.

[53] T.M. Duncan and C. Dybowski, *Surface Sci. Rep.*, **165** (1981) 157.

[54] M. Heming and E. Roduner, *Surface Science*, **189/190** (1987) 535.

[55] M. Heming, *Z. Phys. Chem. Neue Folge*, **151** (1987) 35.

[56] J.P. Stewart, *Quantum Chemistry Program Exchange*, No. 464.

[57] M.J.S. Dewar and W. Thiel, *J. Amer. Chem. Soc.*, **99** (1977) 4899. *J. Amer. Chem. Soc.*, **99** (1977) 4907.

[58] H. Fischer, *J. Chem. Phys.*, **37** (1962) 1094.

[59] J.A. Pople, D.L. Beveridge, and P.A. Dobosh, *J. Amer. Chem. Soc.*, **90** (1968) 4201.

[60] R.V. Lloyd and D.E. Wood, *J. Amer. Chem. Soc.*, **96** (1974) 659.

[61] K. Münger, *Diplome Thesis* (University of Zürich, 1980).

[62] P.M. Morse, *Phys. Rev.*, **34** (1929) 57.

[63] M.J. Ramos and E. Roduner, *unpublished results*.

[64] S.L. Miller, L.C. Aamodt, G. Dousmanis, C.H. Townes, and J. Kraitchman, *J. Chem. Phys.*, **20** (1952) 1112.

[65] S. DiGregorio, M.B. Yim, and D.E. Wood, *J. Amer. Chem. Soc.*, **95** (1973) 8455.

[66] W.A. Pryor, T.H. Lin, J.P. Stanley, and R.W. Henderson, *J. Amer. Chem. Soc.*, **95** (1973) 6993.

[67] H. Fischer, *Z. Naturforsch.*, **20a** (1965) 428.

[68] J.M. Dust and D.R. Arnold, *J. Amer. Chem. Soc.*, **105** (1983) 1221, 6531.

[69] E. Roduner, G.A. Brinkman, and P.W.F. Louwrier, *Chem. Phys.*, **88** (1984) 143.

[70] D. A. Geeson, Ch.J. Rhodes, M.C.R. Symons, S.F.J. Cox, Ch. Scott, and E. Roduner, *Hyperfine Interactions*, **32** (1986) 769.

[71] P. Neta and R.H. Schuler, *J. Phys. Chem.*, **77** (1973) 1368.

[72] R.V. Lloyd and D.E. Wood, *J. Amer. Chem. Soc.*, **96** (1974) 659.

[73] S. Dinçtürk, R.A. Jackson. M. Townson, H.Ağirbaş, N.C. Billingham, and G. March, *J. Chem. Soc. Perkin II*, (1981) 1121.

[74] A. Hudson and J.W.E. Lewis, *Molec. Phys.*, **19** (1970) 241.

[75] K. Münger, *personal communication.*

[76] H.M. McConnell, *J.Chem. Phys.*, **24** (1956) 764.

[77] C.J. Rhodes and E. Roduner, *Tetrahedron Letters*, **29** (1988) 1437.

[78] H.G. Viehe, R. Merenyi, L. Stella, and Z. Janousek, *Z. angew. Chem., Int. Ed. Engl.*, **18** (1979) 917.

[79] H.G. Viehe, Z. Janousek and R. Merényi, editors, *Substituent Effects in Radical Chemistry*, (Reidel, Dordrecht) 1986.)

[80] E.G. Janzen, *Accounts Chem. Res.*, **2** (1969) 279.

[81] L. Sylvander, L. Stella, H.G. Korth, and R. Sustmann, *Tetrahedron Letters*, **26** (1985) 749.

[82] H.-G. Korth, P. Lommes, R. Sustmann, L. Sylvander, and L. Stella, in *Substituent Effects in Radical Chemistry*, Ed. H.G. Viehe et al. (Reidel, Dordrecht) 1986.

[83] Y. Ito, B.W. Ng, Y.C. Jean, and D.C. Walker, *Can. J. Chem.*, **58** (1980) 2395.

[84] P.W.F. Louwrier, G.A. Brinkman, and E. Roduner, *Hyperfine Interactions*, **32** (1986) 831.

[85] G.G. Myasishcheva, Yu.V. Obukhov, V.S. Roganov, and V.G. Firsov, *Khim: Vys. Energii*, **3** (1969) 510 [English transl. High Energy Chem. 3 (1969) 463].

[86] G.G. Myasishcheva, Yu.V. Obukhov, V.S. Roganov, and V.G. Firsov, *Khim: Vys. Energii*, **4** (1970) 447 [English transl. High Energy Chem. 4 (1970) 398].

[87] E. Roduner and D.M. Garner, *unpublished results.*

[88] E. Roduner, *Hyperfine Interactions*, **17-19** (1984)785.

[89] E. Roduner and B.C. Webster, *J. Chem. Soc., Faraday Trans. 1*, **79** (1983) 1939.

[90] Y.C. Jean, B.W. Ng, J.H. Brewer,D.G. Fleming, and D.C. Walker, *J. Phys. Chem.*, **85** (1981) 451.

[91] P.W. Percival, E. Roduner, and H. Fischer, *Chem. Phys.*, **32** (1978) 353.

[92] G. Wikander and O.E. Mogensen, *Chem. Phys.*, **72** (1982) 407.

[93] J.M. Warman, K.D. Asmus, and R.H. Schuler, *Adv. Chem. Series*, **82** (1968) 25.

[94] S.J. Rzad, P.P. Infelta, J.M. Warman, and R.H. Schuler, *J. Chem. Phys.*, **52** (1970) 3971.

[95] P.W. Percival, *J. Chem. Phys.*, **72** (1980) 2901.

[96] E. Roduner *unpublished results*

[97] M. Anbar and E.J. Hart, *J. Am. Chem. Soc.*, **86** (1964) 5633.

[98] D.J. Arsenau, D.M. Garner, M. Senba, and D.G. Fleming, *J. Phys. Chem.*, **88** (1984) 3688.

[99] A. Hill, G. Allen, G. Sterling, and M.C.R. Symons, *J. Chem. Soc., Faraday Trans. 1*, **78** (1982) 2959.

[100] E. Roduner, G.A. Brinkman, and P.W.F. Louwrier, *Hyperfine Interactions*, **17-19** (1984) 803.

[101] P. Neta and R.H. Schuler, *J. Amer. Chem. Soc.*, **94** (1972) 1056.

[102] J.M. Stadlbauer, B.W. Ng, R. Ganti, and D.C. Walker, *J. Amer. Chem. Soc.*, **106** (1984) 3151.

[103] S. Steenken and N.V. Raghavan, *J. Phys. Chem.*, **83** (1979) 3101.

[104] N.V. Raghavan and S. Steenken, *J. Amer. Chem. Soc.*, **102** (1980) 3495.

[105] E. Roduner, G.A. Brinkman, and P.W.F. Louwrier, *Hyperfine Interactions*, **17-19** (1984) 797.

[106] E. Roduner and K. Münger, *Hyperfine Interactions*, **17-19** (1984) 793.

[107] G. Brunton, D. Griller, L.R.C. Barclay, and K.U. Ingold, *J. Amer. Chem. Soc.*, **98** (1976) 6803.

[108] M. Heming and E. Roduner, *Hyperfine Interactions*, **32** (1986) 747.

[109] P. Burkhard, E. Roduner, and H. Fischer, *Int. J. Chem. Kinetics*, **17** (1985) 83.

[110] E. Roduner, W. Strub, P. Burkhard, J. Hochmann, P.W. Percival, H. Fischer, M.J. Ramos, and B.C. Webster, *Chem. Phys.*, **67** (1982) 275.

[111] M.J. Ramos, D. McKenna, B.C. Webster, and E. Roduner, *J. Chem. Soc., Faraday Trans. 1*, **80** (1983) 255.

[112] M.J. Ramos, D. McKenna, B.C. Webster, and E. Roduner, *J. Chem. Soc., Faraday. Trans. 1*, **80** (1983) 267.

[113] W. Strub, E. Roduner, and H. Fischer, *J. Phys. Chem.*, **91** (1987) 4379.

[114] G.A. Brinkman, *Adv. Inorg. Chem. Radiochem.*, **28** (1984) 101.

[115] Ch.J. Rhodes, M.C.R. Symons, C.A. Scott, E. Roduner, and M. Heming, *J. Chem. Soc., Chem. Commun.*, (1987) 448.

[116] S.F.J. Cox, D.A. Geeson, Ch.J. Rhodes, E. Roduner, and M.C.R. Symons, *Hyperfine Interactions*, **32** (1986) 763.

[117] C.J. Rhodes, M.C.R. Symons, and E. Roduner, *J. Chem. Soc., Chem. Commun.*, (1988) 3.

[118] P.W.F. Louwrier, G.A. Brinkman, C.N.M. Bakker, and E. Roduner, *Hyperfine Interactions*, **32** (1986) 753.

[119] D.A. Geeson, M.C.R. Symons, E. Roduner, H. Fischer, and S.F.J. Cox, *Chem. Phys. Letters*, **116** (1985) 186.

[120] Ch.J. Rhodes, M.C.R. Symons, C.A. Scott, E. Roduner, and M. Heming, *J. Chem. Soc., Chem. Commun.*, (1987) 447.

[121] C.J. Rhodes, M.C.R. Symons, E. Roduner, and C.A. Scott, *Chem. Phys. Letters*, **139** (1987) 496.

[122] R.F. Kiefl, M. Celio, T.L. Estle, G.M. Luke, S.R. Kreitzman, J.H. Brewer, D.R. Noakes, E.J. Ansaldo, and K. Nishiyama, *Phys. Rev. Letters*, **58** (1987) 1780.

[123] M. Riccò, C. Bucci, R. De Renzi, G. Guidi, P. Podini, R. Tedesci, and C.A. Scott, *J. Chem. Phys.*, **86** (1987) 4198.

[124] K. Nishiyama, K. Ishida, K. Nagamine, T. Matsuzaki, K. Kuono, H. Shirakawa, R. Kiefl, and J.H. Brewer, *Hyperfine Interactions*, **32** (1986) 551.

[125] E. Roduner and D.M. Garner, *Hyperfine Interactions*, **32** (1986) 733.

[126] T.A. Claxton and A.M. Graham, *J. Chem. Soc., Chem. Commun.*, (1987) 1167.

[127] S.F.J. Cox, T.A. Claxton, and M.C.R. Symons, *Radiat. Phys. Chem.*, **28** (1986) 107.

[128] D. McKenna and B.C. Webster, *J. Chem. Soc., Faraday Trans. 2*, **81** (1985) 225.

[129] J. Manz and J. Römelt, *Nachr. Chem. Tech. Lab.*, **33** (1985) 210.

[130] J. Manz, R. Meyer, and J. Römelt, *Chem. Phys. Letters*, **96** (1983) 607.

[131] F. Jent, H. Paul, E. Roduner, M. Heming, and H. Fischer, *Int. J. Chem. Kinetics*, **18** (1986) 1113.

[132] B. Schrader, J. Pacansky, and U. Pfeiffer, *J. Phys. Chem.*, **88** (1984) 4069.

[133] R. Köster and K.D. Asmus, *Z. Naturforsch.*, **26 b** (1971) 1104.

Index

Lecture Notes in Chemistry